THE
REFERENCE
SHELF

THE PROBLEM

OF WASTE DISPOSAL

edited by ROBERT EMMET LONG

THE REFERENCE SHELF

Volume 60 Number 5

THE H. W. WILSON COMPANY

New York   1989

THE REFERENCE SHELF

The books in this series contain reprints of articles, excerpts from books, and addresses on current issues and social trends in the United States and other countries. There are six separately bound numbers in each volume, all of which are generally published in the same calendar year. One number is a collection of recent speeches; each of the others is devoted to a single subject and gives background information and discussion from various points of view, concluding with a comprehensive bibliography. Books in the series may be purchased individually or on subscription.

Library of Congress Cataloging-in-Publication Data

Main entry under title:

The Problem of waste disposal / edited by Robert Emmet Long.

    p. cm. — (The Reference shelf ; v. 60, no. 5)
    Bibliography: p.
    ISBN 0-8242-0769-6
    1. Factory and trade waste—Environmental aspects—United States. 2. Refuse and refuse disposal—Environmental aspects—United States. I. Long, Robert Emmet. II. Series.
TD897.P75   1989
363.7'28—dc19   88-13825
                                                 CIP

Printed in the United States of America

CONTENTS

PREFACE 5

I. REFUSE AND THE ENVIRONMENT

Editor's Introduction 7
J. Tevere MacFadyen. Where Will All the Garbage Go? .
....Atlantic 8
Tony Davis. Garbage: To Burn or Not to Burn? ....
.............................Technology Review 17
Carolyn Mann. Garbage In, Garbage Out .... Sierra 20
Allen Hershkowitz. Burning Trash: How It Could Work
.............................Technology Review 24
Barbara Goldoftas. Recycling: Coming of Age ....
.............................Technology Review 34

II. OCEAN WASTE DISPOSAL

Editor's Introduction 47
Beverly M. Payton. Blight in the Bight: Sewage and Wa-
ter Don't Mix .... Oceans 48
Anthony Wolff. Fecal Follies .... Audubon 57
Tim Smart and Emily T. Smith. Troubled Waters ......
.... Business Week 62
William Lahey and Michael Connor. The Case for
Ocean Waste Disposal ....Technology Review 70
John W. Farrington, Judith M. Capuzzo, Thomas M.
Leschine, and Michael A. Champ. Ocean Dumping ..
.... Oceanus 84

III. HAZARDOUS WASTE

Editor's Introduction 101

Ed Magnuson. A Problem That Cannot Be Buried .....
....Time   102
Judith Miller and Mark Miller. The Midnight Dumpers. .
.... USA Today   111
James Cook. Risky Business ....Forbes   123
Andrew Porterfield and David Weir. The Export of U.S.
    Toxic Wastes ....Nation   132
James J. Holbrook. Biology's Answer to Toxic Dumps ..
.... Sierra   140
Fred H. Tschirley. Dioxin .... Scientific American   143

IV. NUCLEAR WASTE

Editor's Introduction   153
American Nuclear Society. Nuclear Energy Facts: Ques-
    tions and Answers ....
.... American Nuclear Society Pamphlet   155
Marilynne Robinson. Bad News from Britain .. Harper's   167
Tom Yulsman. Burying Nuclear Waste .. Science Digest   180
Gale Warner. Low-level Lowdown .... Sierra   182
Susan Q. Stranahan. The Deadliest Garbage of All .....
.... Science Digest   188
Cynthia Pollock. The Closing Act: Decommissioning Nu-
    clear Power Plants .... Environment   195

BIBLIOGRAPHY   207

PREFACE

During the 1980s the problem of waste disposal forced itself on the public consciousness. In a time of rapidly accelerating consumerism, the country has heedlessly discarded waste materials and now risks being suffocated or poisoned by them. Although landfills have provided a quick solution to waste disposal, space is now becoming scarce, particularly in the populous Northeast, and new methods are needed. But alternatives, chiefly mass burn incineration, may bring other problems, adding to the pollution of the environment. Underlying the dilemma of waste disposal are simple facts of economics. Pollution reduction for incineration will be expensive; a cleanup of toxic waste sites throughout the country will be staggeringly costly; and better management of America's coastal waters will involve billions of dollars. Where will all this money come from, and if it is obtained, will it be wisely spent?

This compilation addresses all of these important matters. In Section I, various aspects of the problem of sewage and refuse disposal are considered: the shortage of landfill space; the reluctance of communities to serve as sites for incineration plants; the question of whether these plants will be equipped with the latest pollution-reducing technology; and the need to recycle. Section II focuses upon waste, both trash and sewage sludge, that is poured into the sea, befouling harbors and destroying marine life; and the question of how much waste can be dumped safely into the ocean.

Section III deals with hazardous or toxic waste that has been poisoning the environment. Toxic waste sites have frequently been in the headlines since the Love Canal disaster in 1978, and a government effort has been underway since then, financed by the "Superfund," to clean them up. But, as articles in this section illustrate, progress has been faltering and slow, and the bill for cleaning up is very high. The problem of nuclear waste disposal, the subject of Section IV, is equally urgent. In the 1950s, when the first nuclear power plants were built, scientists speculated rather confidently that a solution would eventually be found for storage of the deadly waste of this new industry. But no permanent storage site has yet been selected. The absolute safety of this material in the course of future ages has proved more difficult to verify than was at first thought, and low-level nuclear waste has also been a problem. Leaks have been detected at various landfill sites, and strategies for future storage sites may or may not be well conceived. Clearly all of these aspects of waste disposal are cause for concern and reflection.

The editor is indebted to Ronald Scrudato, research specialist on waste disposal and assistant provost at the State University of New York at Oswego; Dr. Lawrence Abrahamson, of the College of Environmental Science and Forestry at Syracuse University; and Robert Burtch and Debbie Cienava, nuclear information officers for Niagara Mohawk's nuclear power facilities at Oswego, New York. He is also indebted to the authors and publishers who gave permission to reprint the materials in this compilation.

ROBERT EMMET LONG

November 1988

I. REFUSE AND THE ENVIRONMENT

EDITOR'S INTRODUCTION

The widely publicized odyssey of the garbage barge *Mobro* from Islip, Long Island to many distant ports, all of which refused to receive its cargo, and return to its place of origin, was a prominent sign of the problem of waste disposal in the U.S. today. The country, especially in its densely settled northeastern corridor, is running out of landfill space, and a new strategy for waste management is now needed. A huge consumer society, the U.S. has also been a huge disposing society, with an "out of sight, out of mind" psychology and lack of long-range planning. The opening section of this compilation raises questions about what course our large cities and small communities should follow. With landfill space at a premium, they may now incinerate or begin recycling programs, or both; but foresight will be needed to protect the environment.

In the first article in this section, J. Tevere MacFadyen, writing in the *Atlantic*, looks at the dilemma confronting New York City, whose mammoth Fresh Kills landfill is now nearly full. The city must convert to incinerator plants, and do it quickly, but no one wants such facilities in their "backyard"—a problem that also confronts other cities across the land. In a second article, Tony Davis, in *Technology Review*, considers drawbacks of waste-to-energy incineration, especially its emissions of such substances as dioxin, believed by many scientists to be carcinogenic. In an article reprinted from *Sierra*, Carolyn Mann reflects on the pollution caused by mass burn plants.

In the following article, from *Technology Review*, Allen Hershkowitz writes more optimistically of the future for incineration. Pointing to Japanese and European models, he notes that improvements in the incineration of solid waste are possible and could reduce noxious emissions. He also urges recycling, a practice perfected abroad. Recycling is the subject of a final article in this section, that of Barbara Goldoftas in *Technology Review*. Noting that recycling is a growing national trend, she points out that New York State has set a goal to reduce garbage by 50 percent

by 1997, partly through recycling programs. Goldoftas illustrates
how recycling has worked in the U.S. in the past, particularly during World War II, and how it can be reintroduced successfully today.

WHERE WILL ALL THE GARBAGE GO?[1]

The largest landfill in the world sprawls over 3,000 acres at
Fresh Kills, on Staten Island. It is an awesome sight: a sharply corrugated expanse of earth-covered berms built exclusively of compacted waste, a surreal procession of garbage hills and garbage
dales and glittering garbage plains where bulldozers scuttle about
like oversized scavenger beetles, towing trolleys full of trash.
Fresh Kills is almost unimaginably big—big enough, in the conventional measure of large American spaces, to swallow some
2,000 football fields. Its summit, which towers several hundred
feet above sea level, is still rising; 11,000 tons of new garbage ar-
rive every day. Of the approximately 24,000 tons of household
and commercial refuse collected daily by the city, better than
three quarters winds up in New York's three remaining landfills.
One of these accepts only a fraction of the total, and another, at
Fountain Avenue in Brooklyn, is scheduled to close in December,
when, by arrangement with the Department of the Interior, it
will be incorporated into the adjoining Gateway National Recre-
ation Area. For all practical purposes, that will leave only Fresh
Kills.

Paul Casowitz, the deputy commissioner for resource-
recovery and waste-disposal planning at the New York City Department of Sanitation, has the unenviable assignment of figur-
ing out what to do with all the garbage. The prospect of sending
20,000 tons a day to Staten Island is disquieting, but at present
Casowitz has few other options. New York's existing waste-
disposal facilities are quite literally exhausted. The city's landfills

[1]Excerpt of an article by J. Tevere MacFadyen, author and specialist on urban
and environmental affairs. Reprinted by permission of Den Cengden Assoc. Inc.
Copyright © 1985 by J. Tevere MacFadyen. Originally appeared in the March 1985
issue of *Atlantic*.

are either filled or filling fast, and its network of incinerators and handling facilities has been obsolete for some time. Additional landfill space is almost impossible to find, and proposals for high-efficiency incinerators have until recently been mired in the inertia of municipal politics. The city's request to the Department of the Interior for an extension of its deadline was turned down. Landfills, Casowitz remarks hopefully, are like soft luggage: you can always stuff a bit more in. He estimates that using the best available technology he may be able to push Fresh Kills's plateau up to 500 feet, but even the most optimistic analyses suggest that the site will be brim full in twelve to fifteen years. And that will leave no dumping grounds at all. Near the landfill's main entrance, where trucks queue up and sea gulls hover overhead, I found an eerily prophetic sign: DEPARTMENT OF SANITATION, NO DUMPING HERE.

The magnitude of New York's dilemma may be exceptional but the general problem itself is not: though reliable nationwide numbers are hard to come by, cities everywhere seem to be running out of places to put their garbage. "There's no way you're going to get hard figures," says Ronald Musselwhite, who studies the subject for the U.S. Conference of Mayors. "Nobody's even sure how many landfills there are, never mind how much capacity is left. But when enough cities say they have a local problem, you can safely assume the problem is national as well." The extent of the problem is all the harder to measure because municipal solid waste is typically disposed of by a combination of public and private means. And solid waste is considered a purely local concern, whereas sewage, for instance, gets attention from state and federal authorities.

The problem is most keenly felt in older northeastern and central cities, like New York, Boston, Cleveland, Philadelphia, Chicago—densely populated places where the suburban fringe is thickly settled and space is at a premium. In the Southwest and the West open land is cheaper and more plentiful, but landfills are nowhere very popular. Outlying communities, aware of the perils of toxic dumping, are determined to resist dumps of any kind, so municipalities have been forced to expand on sites they already have. Disposal is thus becoming ever more centralized, and the greater the distance between the point of collection and the point of final disposal, the higher the cost. Landfill depletion is not widely discussed in the press, but it is a hot topic among mayors and city managers. The seriousness of the situation is perhaps best indicated by the number of U.S. cities now pursuing waste-disposal alternatives—roughly 200, from Tuscaloosa, Alabama, to Rutland, Vermont.

From the late nineteenth century until the start of the Second World War, New York, like most American cities, maintained a diverse and decentralized waste-disposal system consisting mainly of small incinerators and landfills. Many of the incinerators were abandoned when they came to the end of their useful life and were never replaced, leaving landfills to take up the slack. Planners acknowledged that landfill capacity would eventually be exhausted but saw no immediate reason for concern. As additional incinerators were retired and smaller landfills closed, the system contracted. New York in 1934 counted twenty-two working incinerators and eighty-nine landfills. In 1984 only three of each remained.

Ultimately, Paul Casowitz explains, virtually all garbage is disposed of in landfills. Bulk can be reduced by baling, burning, shredding, or other, more exotic techniques, but while these extend the life of a landfill, they do not obviate the need for one. Contrary to popular belief, ocean dumping off New York no longer occurs, and it hasn't occurred since 1934. The heavily laden barges that skirt the Statue of Liberty nowadays are en route to Fresh Kills. Open burning of dumps, prevalent in rural regions until the late 1960s, was all but eliminated by stricter air-quality standards. Many older incinerators, lacking adequate pollution controls, either were closed or had their operations sharply curtailed. With more and more of the waste stream directed at landfills, the land naturally filled faster. Rapid expansion and heavier loads brought new problems: methane-gas buildup, groundwater contamination from chemical leachates, uneven settling and instability. By the early seventies most municipal authorities who cared to look could see trouble ahead.

At about that time the burgeoning ecology movement intersected with the first energy crisis to produce a surge of interest in what came to be known as "urban ore." Garbage, some suggested, was too precious to throw out. It contained all manner of valuable resources, which, if they could only be separated from the rest of the waste, might profitably be sold. A minor boom ensued. An assortment of unproved and often highly speculative resource-recovery technologies were proposed, the majority falling under the general heading of "refuse-derived fuels." The RDF concept seemed beguilingly simple. Once the nonburnable recyclables (chiefly glass and ferrous metals) had been removed, the remainder of the waste would be mechanically reconstituted into a storable, transportable, and above all salable fuel that could be burned to produce energy. RDF, unlike landfills, was a politically attractive idea. Bonds were floated and pilot projects built, and all across America mayors and municipal officials breathed sighs of relief.

The initial prognosis for refuse-derived fuels sounded too good to be true, and, unfortunately, it was. The first wave of RDF facilities was plagued by problems. Equipment failed to function properly. Markets failed to materialize. Mechanical problems were legion: most of the systems relied on high-horsepower slicing and dicing of the waste, and this proved to be more difficult than had been anticipated. "When you're sorting something as heterogeneous as garbage," says Ben Miller, a planner on Paul Casowitz's staff, "a pair of pantyhose can shut the whole thing down, as happened more than once. And there's just the regular abrasive effect of all that gritty stuff, tearing up your conveyor belts and wearing out your elbow joints and that sort of thing." Dust and emissions problems continually occurred, along with occasional explosions. In some places—Bridgeport, Chicago, and Baltimore, for example—RDF plants opened and closed without ever operating satisfactorily. Other cities, mostly smaller communities with municipally owned utilities, deemed their investments in RDF projects too great to write off. Some facilities worked efficiently, but too few did, and enthusiasm rapidly waned in both public and private sectors. "Meanwhile," Ben Miller says, "ten years went by during which basically nothing was done to reduce the rate of landfill depletion."

"What happened," Miller says, "was that the activity in the early seventies was spurred not so much by concern about diminishing landfill capacities as by an interest in possible profits." Paul Casowitz puts it this way: "We forgot what the point was. You don't get into the business of resource recovery because you think there's gold in the garbage. It's just not there. The point is getting rid of the garbage, and getting rid of garbage is going to be expensive no matter what technology you use. It's a strategic error to view resource recovery as an economic-development scheme. If economic development becomes your driving force, you aren't going to solve the waste-disposal problem." The early RDF experiments may actually have been counterproductive in that they left business and government apprehensive about alternative waste-management plans and effectively postponed the inevitable confrontation with unmanageable quantities of garbage. "You can't recover lost time," Casowitz says. "The longer you wait, the tighter the noose gets."

In 1981 the New York City sanitation commissioner, Norman Steisel, accepted contract bids and formally proposed that the city construct a "mass-burning," waste-to-energy resource-recovery facility at the old Brooklyn Navy Yard, in the Williamsburg neighborhood of Brooklyn. This, he warned, would be only the first of many such plants that would have to be built within the next decade if New York was to solve its waste-disposal problem.

Mass burning can hardly be called experimental. In contrast with refuse-derived fuels, the technology seems direct and uncomplicated. In a mass-burning plant nothing is sorted or pre-screened, though some ferrous metals may be recovered. Undifferentiated waste is incinerated at high temperatures in a furnace lined with water-filled tubes. (Hence the term "waterwall" is often applied to the process.) Heat from the burning garbage generates steam, which may be used as is, fed into a municipal steam system, sold, or made to turn turbines, where its energy is converted into electricity. Most of what doesn't burn is hauled to the landfill. Mass burning is neither as sophisticated nor as innovative as RDF, but it can reduce the volume of waste by as much as 90 percent, and it is both proven and reliable. In Europe, where the landfill-depletion problem became evident much earlier on, mass-burning facilities have been used for more than forty years.

At present mass burning is the technology of choice among waste-disposal planners like Steisel and Casowitz, who see their mission in terms not of potential profit but of landfill reduction and getting rid of the garbage at the lowest possible cost. Many are veterans of RDF debacles and are cautious about the long-term consequences of any plan. In an era of shrinking municipal governments the trend is toward shedding old responsibilities rather than taking on new ones, and the majority of cities would love to get out of the waste-disposal business. Although the private sector is now skeptical of RDF, it has pursued the notion of mass burning with enthusiasm, and most proposed mass-burning installations—including New York's—have been designed with the expectation that they will be either municipally owned and privately operated or wholly independent projects.

Suffice it to say that Steisel's mass-burning proposal was not warmly received. "When I started this job," Paul Casowitz says, "I thought I was getting involved in a motherhood-and-apple-pie issue. It seemed to me that turning garbage into energy was something everybody had to support." He soon learned otherwise. Williamsburg residents, to the surprise of absolutely nobody, mobilized against the plan. Environmentalists complained about potential air pollution from particulate emissions, acid gases, and trace toxins, and protested that not enough effort had gone into recycling programs aimed at trimming the size of the city's waste stream. City politicians, even as they endorsed the idea in principle, deftly avoided any commitment to its realization.

Three years went by. Steisel and Casowitz regrouped. They held public hearings. They lobbied and cajoled. Hoping to overcome the political obstacle (waste-disposal planners call it the Not-In-My-Back-Yard Syndrome), they drafted a revised and more comprehensive plan. This time they proposed seven new plants in addition to the Navy Yard facility, to be located in all five of the city's boroughs. Steisel referred to this as a "share the pain" plan, and he hoped it would help to neutralize at least some of the opposition. In pleading his case once again, in April of 1984, the Commissioner said, "I am concerned that we have already lost too much time. It currently takes six to eight years to implement capital projects such as new schools and firehouses, which often have considerable public support. . . . There is no time left to defer the difficult decisions that must be made, without jeopardizing the welfare of the city as a whole." The Board of Estimate, New York's fiscal politburo, postponed action until further studies could be made.

Finally, in late December of last year, after marathon arm-twisting by the mayor, a narrow majority on the board approved continued planning for five facilities. Though that move for the first time makes an incinerator program part of city policy, it remains to be seen how much real action will result, and how quickly. The board's vote constitutes a nonbinding recommendation, and there is no guarantee that the board will not later reverse itself, as it has in the past. Public hearings, environmental-impact

statements, and city and state approvals are still required before
any plant can be built. The only site for which this lengthy process has even been initiated is the Navy Yard, but Howard Gol-
den, the powerful Brooklyn borough president, has gone on
record as vigorously opposed to construction at that location, and
few observers doubt his ability to slow things down. The Department of Sanitation says it hopes to have mass-burning facilities
operating by 1991, but for now that timetable seems rooted more
in optimism than in reality.

The oldest mass-burning waterwall waste-disposal plant in the
United States, and one of the largest, is the Chicago Northwest
Waste-to-Energy Facility. It has operated constantly since September of 1970, accepting an average of 1,000 tons of garbage
a day and producing steam for sale. Its essential technology, like
that of all U.S. mass-burning facilities, is European, in this case
developed by the Josef Martin Company, of West Germany. The
plant is located in a hardscrabble mixed-use neighborhood where
older housing backs up against heavy industry, and it fits right in.
John Ellis, the plant manager, offered to show me around.

"The only consistent thing about garbage is its inconsistency,"
Ellis remarked. "It's probably the most heterogeneous material
available." We were standing on a narrow balcony overlooking
the tipping floor, where trucks discharge garbage into a 10,000-
cubic-yard storage pit. Behind us were the gaping maws of the
plant's four furnaces. High above, a crane operator maneuvered
his machine's cumbrous jaws, depositing huge mouthfuls of mis-
cellaneous refuse into the furnace's feed hoppers: crimson velour
sofa cushions and a plastic-wicker laundry basket, two automobile
tires and the flattened, two-dimensional frame of a child's tricy-
cle. Ellis explained that the operator wasn't just shoveling stuff
in but was trying to alternate wet and dry to achieve the perfect
mix. "Fuel management," Ellis called it. Check stubs, one brown
sock, the label off a bottle of Canoe cologne, lots of grass clip-
pings. "Grass is high-cellulose. Very moist. But the real problem
is that it attracts dirt, which doesn't burn, of course, and causes
abrasive problems on the grates." Incredibly, the plant was almost
odor-free. "It doesn't stay here long enough to smell," said Ellis,
who noted that the building is negative-pressurized to keep odors
inside. Another truck grumbled up to disgorge its contents.
"Every once in a while we'll get a bad load," Ellis said. "Bedsprings

or construction debris, something like that." He shook his head. "I don't like it. Don't want it. It has no redeeming thermal value."

The garbage falls through the hoppers onto steeply inclined reciprocating grates, where it burns at approximately 1600°F. The steam thus generated is used for various purposes on site or sold, to—of all things—a candy company, which uses it in the manufacture of sourballs and peppermints and cream-filled caramels. "Our steam has to be FDA pure," Ellis said, plainly proud that it is. Residues and ash from the furnaces go to a landfill; ferrous metals currently go unreclaimed. Particulate emissions are mitigated with conventional electrostatic precipitators. Stack exhausts are monitored for acid gases and trace toxins, but to date no special provisions have been made for the control of these substances. In Germany, where rigorous controls are mandated by law, pollution-abatement technologies are the rule. Paul Casowitz sees no reason why similar equipment couldn't be installed at U.S. plants as well, provided we're willing to pay the price. "It's just a question of political will," he says.

The most striking thing about the Chicago Northwest Waste-to-Energy Facility is its ordinariness. A mass-burning plant isn't a playground, obviously, but neither is it the calamity that critics seem to fear, and it is surely no more noxious than a Fresh Kills. The Board of Estimate instructed Commissioner Steisel to respond to concerns about dioxin. His report, released last September, acknowledged that the incidence of cancer might rise by 0.24 to 5.9 cases per million people exposed for seventy years at the point of maximum concentration. The existing cancer rate was pegged at 250,000 cases per million. Barry Commoner, for one, was unconvinced. "We don't understand this thing well enough," he was quoted as telling *The New York Times*. In the same article, Adam Stern, of the Natural Resources Defense Council, sounded slightly more approving. "It has to have proper monitoring," he cautioned. He went on to say, "We don't want to be seen as opponents or trying to slow up the process. We've got to keep moving on this. We've got a real crisis on our hands." Some months later Commoner softened his position, pledging to withdraw his objections if the city would promise to close any incinerator whose dioxin emissions exceeded strict standards.

What are the objections to mass-burning projects? "Well, there are certainly substantive reasons for not wanting one in your neighborhood," says Jim Meyer, a planner for the New York

City Department of Sanitation. Dioxin or no dioxin, mass-burning facilities are heavy industry. A mass-burning plant brings an increase in truck traffic, if nothing else, and some measurable degree of environmental degradation. So does an auto plant, of course, or a steel mill. A more fundamental objection, the one at the heart of the environmental movement's lack of enthusiasm for mass burning, is that a rush to build new incinerators will reduce society's incentive to waste less, to recycle more, and to develop better and more reliable resource-recovery technologies. The real solution to the waste-disposal problem, according to these critics, is to dispose of less waste. Some also argue that new and better technologies, when (and if) they come along, will be difficult to incorporate into mass-burning systems.

"But the next question has to be, What's the alternative?" Ben Miller insists. Garbage disposal is like prisons—a disagreeable social necessity. Everyone realizes that felons have to be incarcerated and trash has to be dumped, but nobody wants a waste-disposal facility or a prison near his or her community. A recently inaugurated mass-burning plant in Westchester County, New York, took twelve years from inception to completion, with the bulk of that time spent seeking a site. As long as the garbage disappears off the curb regularly, the prevailing public sentiment seems to be "out of sight, out of mind"; and the protestations of someone like Paul Casowitz are accorded about as much credence as were the alarms of Chicken Little.

Landfill depletion will probably never be a popular issue. Every politician who has ever run for city office clamors for cleaner streets, but few show much concern for where the sweepings are deposited. New York's constituency for resource-recovery development, Jim Meyer says, lives next door to Fresh Kills. When I asked Paul Casowitz what will happen if, in effect, nothing happens, he replied with weary sarcasm, "Well, there's always Central Park." He went on to amplify slightly. "Is one day going to come when we run out of room? No, probably not. People won't choke on their garbage. But the options will become fewer and the solutions more expensive." As Norman Steisel put it, in the introduction to his most recent proposal:

A collapse of sanitation services due to the total exhaustion of existing landfill capacity is unthinkable and, therefore, unlikely to occur. However, we feel strongly that to defer the implementation of resource recovery will only limit the City's future options, and the need to open new landfills could be one of the undesirable consequences. The impacts of such a scenario were examined as part of this report, but I am sure you will agree that none of us would want to see them become a reality. Such a drastic action, however, might entail reopening a landfill at Ferry Point Park and creating a new one at a site near Co-op City, both in the Bronx, as well as a major expansion of the Fresh Kills site on Staten Island. Even that would only serve to delay the inevitable crisis, leaving future generations with a new and larger disposal gap in only a few short years. At that point, having consumed all remaining landfill capacity, how the City would handle its garbage can only be left to the imagination.

"We're very conservative here," Paul Casowitz says, "or maybe very cynical. Ultimately, as I like to say, the garbage always wins. That's the nice thing about sanitation. It's a real problem. You can talk and talk and preach and posture and throw up all kinds of excuses as to why you don't want to take any action, but eventually you're going to be confronted by the compelling force of reality." . . .

GARBAGE: TO BURN OR NOT TO BURN?[2]

Americans generate 220 million tons of garbage annually, and disposing of it is an immense problem. Since the early 1960s, most communities have dumped their trash into massive landfills. But as the landfills run out of space and neighborhoods raise concerns about threats to the environment, finding new sites becomes virtually impossible.

One proposed alternative is recycling. Dan Saltzman, an Oregon engineer, told the American Institute of Chemical Engineers that although recycling was "once ridiculed as an ineffective hobby of environmentalists," it "is now regarded as an essential component of solid-waste management and a cost-effective way to reduce dependence on landfills."

But many local officials who deal with solid waste doubt that Americans will shed their throwaway habits quickly enough to make recycling a feasible solution. As a result, municipalities are rapidly turning to modern incinerators that turn garbage into energy.

[2]Reprint of an article by Tony Davis, *Technology Review* staffwriter. Reprinted by permission from *Technology Review*, 90:19. F./Mr. '87. Copyright © 1987 by *Technology Review*.

According to the National Municipal Solid Waste Association, 65 energy-recovery plants have gone into operation since 1970. Charles Johnson, technical director of the association, says that 70 more plants are under construction or on the drawing boards. Virtually every major U.S. city is considering building at least one. Johnson predicts that by the end of the century incinerators will be burning 30 percent of the nation's trash, compared with 6 percent today.

Energy-recovery plants are massive. Furnaces can reach six stories high, and emission stacks tower up to 500 feet. Turning trash into energy costs a bundle—$150 million for a plant burning 1,500 tons a day.

A Controversial Solution

However, many environmentalists object to incinerators as well as to landfills. The opposition stems mainly from one word: dioxin. In the late 1970s, studies in the United States, Japan, Switzerland, the Netherlands, Germany, and Canada found that incinerators emit dioxin gases, believed by many scientists to be carcinogenic.

The degree of danger from dioxin emissions is widely debated. Kay Jones, a Seattle-based environmental engineer, says emissions data from 18 plants around the world indicate that only one extra cancer case would result for every 10 million people. In 1985 the New York State Department of Health concluded that an incinerator in Peekskill could produce one to two cancer cases in a million people over 70 years—and that an incinerator in Niagara Falls could produce eleven to twenty.

Ecologist Barry Commoner, director of the Center for the Biology of Natural Systems, disputes the state's calculations. He says the numbers are 17 per million people in Peekskill and 270 per million in Niagara Falls. He cites 1984 research led by Donald Hay, director of urban activities for the government agency Environment Canada. Hay's group measured slight dioxin concentrations in the furnace of an energy-recovery facility in Prince Edward Island, Canada, but a much higher level of dioxin was escaping from the stack.

Advocates of energy recovery insist that incinerators can be made safe. "Emissions can be handled by conventional air-pollution-control equipment," a 1986 American Institute of

Chemical Engineers study states. The study does not mention dioxin, but some research indicates that burning garbage at high enough temperatures will eliminate or drastically reduce dioxin emissions.

The 1984 Canadian research casts doubt on that theory, but in the summer of 1985, Environment Canada scientists suggested a new way to decrease dioxin emissions. They said that if plant operators held the temperature of exhaust gases steady, dioxin would attach to particles less than four-hundredths of an inch in diameter. Emissions could then be controlled with "baghouses"—tall racks with three to five large bag filters hanging down to collect the particles.

Sweden burns half its waste, but in 1984 the country's Environmental Protection Board put a moratorium on new incinerators after high levels of dioxin turned up in crabmeat and mother's milk. In mid-1986, the government lifted the moratorium in favor of a recommended emissions standard for new incinerators of a tenth of a billionth of a gram of dioxin per cubic meter of air.

Per Stor Hammar, a senior technical officer for the agency, is confident that new incinerators can meet the standard. He points to two Stockholm plants that have already done so. First they run stack gases through an electric filter to remove dust. Then they mix the gases with water and use scrubbers to wash out dioxins and heavy metals. Finally, they dry and remove the leftover sludge.

The United States has no dioxin standard, and EPA is conducting studies to determine whether one is necessary. Emission levels at U.S. incinerators can reach as high as 40 billionths of a gram, but they are as low as 2 billionths of a gram at the newest plants.

GARBAGE IN, GARBAGE OUT[3]

When Capt. Duffy St. Pierre pulled his tug *Break of Dawn* away from a Long Island dock on March 22, he thought he was taking a short trip to North Carolina with 3,186 tons of garbage in tow. But after six states and three foreign countries refused to welcome the barge, St. Pierre's cargo became a fly-infested symbol of our throwaway society's disposal problems.

After more than two months at sea, the barge finally returned to New York—where it sits while the cargo's owner, Alabama businessman Lowell Harrelson, waits for court approval to unload the trash and burn it in a Brooklyn municipal incinerator.

Like Harrelson, many U.S. cities are eyeing incinerators as they find themselves increasingly turned away from landfill sites. Americans toss out at least 150 million tons of trash every year, and 90 percent of that winds up buried. An April 1987 Worldwatch Institute study of the world's growing garbage glut revealed that by 1990 half the cities in the United States will have exhausted their landfills. As reports of groundwater contamination from buried garbage grow, city officials across the nation are encountering heavy local opposition to the expansion of existing landfills and to the opening of new ones. Cities often must resort to trucking their trash to rural areas, or even to other states.

It's no wonder, then, that waste-to-energy plants seem like the ideal solution. The plants burn municipal waste as fuel to generate steam and electricity, while reducing the volume of trash by 60 to 90 percent. At least 200 of these facilities are now being planned, built, or operated nationwide. Most are "mass burn" plants, which burn waste without first separating its components.

But along with electricity the plants are generating concern among environmentalists, who say the ash the plants produce and the emissions from their stacks are serious—and virtually unregulated—health hazards. Environmentalists also worry that efforts to reduce waste and to create or expand recycling programs will go up in smoke along with the trash.

[3]Reprint of an article by Carolyn Mann, *Sierra* copy editor. Reprinted by permission from *Sierra*, 72:20–27. S./O. '87. Copyright © 1987 by the author.

The issue of incinerator ash—in particular, how to dispose of it—has drifted to the forefront of the burn-plant debate. In EPA tests, every sample of fly ash, the fine particulate matter trapped in the plants' air-pollution control devices, showed unacceptable levels of toxic metals such as lead and cadmium. Tests of bottom ash, the unburnt residue that collects on an incinerator's grate, showed unacceptable levels of these elements in 10 to 30 percent of the test cases. Concentrations of the potent carcinogens dioxin and furan are also present in fly ash.

In short, incinerators turn bulky garbage into compact, toxic waste. While one would expect to find ash disposal strictly controlled, this isn't the case. Under the Resource Conservation and Recovery Act, it's up to plant operators to identify their hazardous materials. Because ash content varies from day to day, testing must be continuous. Yet operators aren't eager to adopt costly control devices and monitoring programs, and the EPA does not force compliance.

As a result, ash is often dumped in municipal landfills, where its toxic components can leach into groundwater. According to Environmental Defense Fund scientist Richard Denison, many facilities routinely combine fly ash with the less-toxic bottom ash to avoid exceeding established hazard levels. But even the combined ash is failing many tests, Denison reports.

"If the incinerators had to dispose of their ash as hazardous waste," says Cynthia Pollock of the Worldwatch Institute, "it would make the plants ten times more expensive [to operate]." In 1985, according to the *1986–87 Resource Recovery Yearbook*, disposal fees for non-hazardous ash averaged $13 per ton, while those for hazardous ash ran as much as $200 a ton.

The death of federal regulations of the plants' smokestack emissions is another concern. Depending on the sophistication of pollution-control devices used, an incinerator may emit gases that contribute to acid rain, as well as up to 27 heavy metals and extremely toxic dioxin and furan compounds.

Like ash, airborne pollutants (other than solid-particle emissions) are subject to precious little federal regulation: of the 27 toxic metals that incinerators may produce, only three (lead, mercury, and beryllium) are controlled. Increasingly, states are requiring burn-plant builders to install scrubbers that must curtail up to 95 percent of these acid-rain-causing emissions.

Despite their known toxicity, dioxin and furan are not regulated by the federal government. "These are two of the most toxic substances made by man," says Dr. Paul Connett, a chemist at St. Lawrence University in Canton, N.Y. "They can damage the lymphatic system, cause birth defects, and promote cancer." When burned, chlorine compounds in waste products such as plastic, bleached paper, and table salt regroup and form these toxic molecules.

In general, burn-plant emissions do not bode well for air quality. Poor air is already strangling many parts of the country, and the rush to build waste-to-energy plants is likely to exacerbate the problem. "Burn plants can add as much lead to the atmosphere as has been removed by de-leading gasoline," says Sierra Club lobbyist Blake Early. "Communities with high lead levels have no business considering mass burn." Lead's fine particles lodge easily in the lungs, then find their way into the bloodstream and accumulate in bone marrow.

Beyond the debate over incinerator ash and emissions, activists feel that incinerators will do nothing to conserve resources and will gut the recycling programs that have taken years to establish. Recycling advocates point to successful European and Japanese operations, where as much as 65 percent of municipal wastes are recycled, greatly reducing the amount of garbage to be buried or burned. Japan, for example, recycles 95 percent of its beer bottles and two-liter sake bottles. Tossed in incinerators, these noncombustibles simple increase the volume of contaminated ash that must be buried.

Recycling is also cheaper than either dumping or burning, advocates add. The one-ton bales of rotting paper on St. Pierre's barge, for example, might have fetched up to $20 a ton from recyclers but would have cost at least $40 a ton to dump—if a landfill had been willing to accept them. Cities should burn trash only as a last resort to their garbage crisis, activists say—and only after less-damaging waste reduction, recycling, and source-separation programs have been implemented.

"Should incinerators be built before a recycling program is in place, reducing the volume of garbage could be disastrous for plant operators and create conflicts with recycling proponents," wrote Allen Hershkowitz, director of solid-waste research for the New York environmental-research group INFORM, in a recent edition of *Technology Review*. Hershkowitz added that when burn-plant operators are forced to share their garbage with local recyclers after a plant has been designed, they usually seek permission to truck in garbage from other cities—an option most communities fight.

In the meantime, environmental groups, including the Sierra Club, are working to strengthen federal regulation of these plants. Because of a settlement reached with the Natural Resources Defense Council, in July the EPA announced regulations that would require new incinerators to employ the "best available technology" (BAT). That would mean the installation of scrubbers to reduce emissions. Operators would be required to ensure that the largest possible amount of waste and pollutants is destroyed in the burning process. The EPA plans to propose guidelines for reducing emissions from existing incinerators by late 1989.

Environmentalists and legislators are angry, though, that the EPA did not set specific limits on the amount of pollution incinerators may discharge. "The EPA's own data say that these incinerators are sources of numerous carcinogens," said Rep. Henry Waxman (D-Calif.), chair of the House Subcommittee on Health and the Environment, in a *New York Times* interview. "It is an outrage to say they should not be regulated as hazardous air pollutants."

A bill sponsored by Rep. James Florio (D-N.J.), H.R. 2787, is currently struggling through congressional committees. Besides requiring BAT standards for new burn plants, the legislation would set monitoring, operating, and maintenance requirements to ensure that burn-plant operators comply with the Clean Air Act. Florio has also introduced a companion bill, H.R. 2517, to regulate incinerator ash.

Meanwhile, a number of states either have passed or are considering mandatory recycling programs. Rhode Island, New Jersey, and Oregon all recently adopted recycling legislation, and New York has a solid-waste-management plan in the works that aims at recycling 50 percent of its waste within ten years.

But at presstime, 3,186 tons of New York's garbage sits anchored off Coney Island. The load, largely composed of once-recyclable corrugated paper now ruined by sea water, waits to be burned in a scrubberless incinerator built 26 years ago. While in-

cinerating the trash may rid America of a national embarrass-
ment, many environmentalists feel that the real issue has simply
been swept under the rug.

BURNING TRASH: HOW IT COULD WORK[4]

Cities and counties in at least 40 states are operating, building, or considering resource-recovery plants to burn their trash.
These facilities create steam that can be used to heat or cool
buildings or to generate electricity. However, concern about this
technology's potential to cause serious environmental problems
is high. Garbage burning generates a range of pollutants, including gases that contain heavy metals and dioxins and that contribute to acid rain. Moreover, incinerators require a new breed of
secure landfills that accept only the toxic ash they generate.
Thus, citizen groups throughout the United States are battling
local officials in an effort to prevent them from building such
plants.

After conducting a two-year study of trash-burning plants in
Norway, Sweden, West Germany, Switzerland, and Japan, I am
convinced that these technologies can be highly efficient in reducing emissions if the waste is properly prepared and pollution-
control devices are run by highly skilled workers. However, neither the U.S. government nor individual states have established
the full range of standards to guarantee that these steps are taken.
The result is that U.S. operators rarely install the needed devices
or perform the procedures necessary to ensure the best performance. In fact, plant operators usually consider citizen concern
about pollutants a public-relations problem rather than a serious
call for reform. Furthermore, plants known to work poorly are
not shut down. This situation must be remedied if the United
States is to pursue this technology, which is becoming an essential
component in dealing with the crisis in municipal waste.

[4]Reprint of an article by Allen Hershkowitz, author and director of solid-waste
research for INFORM, a non-profit environment-research group in New York. Reprinted by permission from *Technology Review*, 90:26–34. Jl. '87. Copyright © 1987
by *Technology Review*.

U.S. cities and towns generate more than 410,000 tons of solid waste each day. Landfills, until recently the method of choice for disposing of this waste, are closing. New York City has shut 14 landfills over the past 20 years as they have become full and their hazards to public health have become clear. The Fresh Kills landfill on Staten Island, the world's largest and one of only two remaining in the city, violates state and federal law by polluting the nearby Arthur Kill waterway with 4 million gallons of toxic liquid each day. Yet this dump will soon have to accept virtually all the 25,000 tons of solid waste that New York generates each day.

The situation is similar elsewhere. New Jersey officials and industry representatives view the imminent closing of 2 of the state's 13 remaining landfills as a crisis of "unprecedented proportions." All of Seattle's landfills will soon be closed. Michigan has determined that leaching landfills are probably responsible for at least 139 cases of groundwater contamination. The executive director of Mississippi's Department of Natural Resources says that "it will be increasingly difficult, if not impossible, to site a landfill" because of citizen opposition and the threat of groundwater pollution. Only half of the nation's 9,244 municipal landfills have valid operating permits. Communities are beginning to realize that they must take extraordinary precautions to build better landfills, including installing impermeable liners and wastewater collection and treatment facilities. These requirements often make the dumps extremely expensive. Incinerators will not eliminate the need for such landfills, but they can greatly reduce the number required.

The Recycling Option

Many U.S. environmental groups focus on recycling as the solution to the landfill problem. Recycling is already a key component in European and Japanese efforts to handle municipal waste. Yet experience in Japan, which has the world's most successful recycling program, suggests that recycling can take care of 65 percent of municipal solid waste at best.

Machida City, a town with a population of 320,000 about 40 miles south of Tokyo, provides an example of the diligence of Japanese recycling efforts. City officials, including sanitation workers, go door to door at least once a year explaining the purpose of separating wastes, and the municipality distributes brochures about the benefits of recycling to third-grade and fourth-grade children. Like most of Japan's 3,255 municipalities, Machida residents separate their waste into seven general categories: newspapers, combustibles (including organic kitchen wastes, light plastics, and soiled paper), non-combustibles (hard plastics, broken glass, and scrap metal), glass bottles, aluminum and steel cans, hazardous materials (including batteries and other items containing mercury or cadmium), and bulky wastes (such as furniture and bicycles).

Over 100 civic groups collect residents' newspapers, glass bottles, and metal cans, and the sale of these items provides income to support their activities. Over 95 percent of all newspapers, 50 percent of glass bottles, and over 70 percent of steel and aluminum cans are recovered. In a program known as *chirigami kokan* (tissue-paper exchange), citizens receive weekly allotments of tissue paper, napkins, and toilet paper in return for their week's worth of newspapers. The town collects hazardous household materials at a citizen's request and stores them on Hokkaido, Japan's northernmost island, until methods for recycling this waste can be found. The town also collects bulky items once a month and sends them to a "recycling cultural center" where handicapped citizens repair them for resale. The center also composts some kitchen wastes for use in its greenhouse. Repaired bulky items and flowers from the greenhouse are sold to help support the center's rehabilitation program. Non-combustibles are sent to the local landfill.

Such an ambitious program is extraordinary by any standards. Yet Machida City must still burn about 1,250 tons of non-recycled waste—about 34 percent of the total—each week. The city does this in an incinerator that supplies steam to heat the cultural center, including its greenhouse and swimming pool.

Thus, even the Japanese have found total recycling impossible. Two key reasons are the heterogeneous nature of municipal waste and the limited market for recycled goods. For example, polyvinyl chloride (PVC), common resin, has to be separated from other plastic resins such as polyethylene or polystyrene before it can be melted and reused. It can simply be shredded and used as a filler in clothing, but the market for such uses is limited.

As many as 8,000 U.S. communities are recycling one or more commodities, but reductions in waste levels are nowhere near as high as in Japan. Some of the best programs—including those of Davis, Calif., and Camden, N.J., where as many as three of four citizens participate—have reduced waste levels by no more than 25 to 30 percent. Of course, the United States has not made recycling a priority, as indicated by the federal subsidies provided to haulers of virgin materials, and the fact that the government does not monitor the percentage of recyclable commodities that are reused. Still, growing numbers of states and towns are pursuing recycling as an important option in controlling waste, and Oregon, Rhode Island, and New Jersey have made recycling of some wastes mandatory.

The United States could also do more to reduce the amount of waste it generates. The per capita production of garbage is significantly less in Europe and Japan than in this country. The average resident of Oslo, Norway, generates about 1.7 pounds of garbage a day while the average Japanese, Swedish, West German, and Swiss citizen produces about 2.5 pounds a day—a sharp contrast with the U.S. average of 4 to 6 pounds a day. However, efforts to restrict excess packaging, say, even when combined with recycling, cannot eliminate the disposal problem entirely. Communities are finding that they must look to incineration to dispose of at least some of their waste.

The Economics of Incineration

About 70 resource-recovery plants are already in operation or undergoing start-up testing in the United States, 20 more are under construction, and over 100 are planned. A small plant that burns 50 tons per day, such as that being built in Batesville, Ark., can cost $1.2 million, while $570 million is being spent for the 4,500-ton-a-day complex under construction in Broward County, Fla. New York City's Department of Sanitation has estimated that building 8 waste-to-energy plants capable of burning 70 percent of the city's daily garbage will cost $3 billion.

This is a significant investment for most communities, but the sale of energy recovered from the incineration process offsets some of the burden. The garbage is burned in a funnel-shaped furnace, and a boiler within or above the furnace produces steam from the hot combustion gases. This steam can be used directly to heat buildings, run air conditioners, or power the turbine of a generator, which makes electricity that is then sold to a utility.

Electricity production can range from 11 megawatts from a plant that burns 550 tons of refuse per day to over 100 megawatts from a plant the size of Broward County's.

Municipalities usually contract with manufacturers of resource-recovery plants to build them. The communities may then run the plants themselves or—more commonly—have the manufacturer or another company run them. Such a venture can be quite profitable. The waste-to-energy industry is unique in that operators receive income from both the fuel they burn—towns pay a "tipping fee" for each ton of waste they deposit—and the energy they generate. A plant's manufacturer also sees a 20 to 35 percent pre-tax return on the construction of the facility. In return a community may strike a deal on the use of the energy from the plant. For example, Peekskill, N.Y., receives a $1 million yearly credit from Con Edison, which buys the electricity produced by the town's incinerator.

Resource-recovery plants generate energy much less efficiently than other alternatives such as cogeneration plants. Yet it would be a mistake for towns to regard their incinerators as energy-producing ventures. Rather, they are an important means of disposing of municipal waste.

Recycling Makes Burning Better

Although recycling cannot solve the entire solid-waste disposal problem, it is essential to any incineration scheme. High-volume incinerators are usually a community's most complex investment, and the savings from reducing the amount of waste to be burned can be significant. Reducing plant scale and cost can also shorten the inevitable siting struggle associated with incinerators. Conversely, should incinerators be built before a recycling program is in place, reducing the volume of garbage could be financially disastrous for plant operators and create conflicts with recycling proponents.

Communities that plan incinerators before they adopt a recycling program often find that plant operators will seek permission to important garbage from other areas. This can also exacerbate siting struggles, as shown in Hempstead, N.Y., New Haven, Conn., and Collier County, Fla. Citizens have enough difficulty accepting an incinerator to burn their own community's waste; they often refuse to accept a facility that burns another community's wastes.

Recycling can also be essential in improving the efficiency of an incinerator. Although bottles and metal cans are not combustible, most U.S. communities routinely throw them into the waste stream along with everything else. These materials interfere with good burning and wind up as contaminated ash in the bottom of the incinerator, which must be dumped into landfills. This increases the amount of maintenance a plant requires, depletes scarce landfill space, and inflates disposal costs, which can approach $100 per ton for incinerator ash. The Japanese, who separate most bottles and cans from the rest of their garbage before burning, are amazed that Americans do not do this. U.S. plants generate more than twice the amount of bottom ash that Japanese plants do—10 percent of the waste's original volume.

Burning metals can also create serious health problems. Resource-recovery plants can give off as many as 27 metals, including antimony, arsenic, beryllium, cadmium, chromium, lead, mercury, nickel, tin, and zinc. Except for beryllium, lead, and mercury, these substances are not covered by U.S. federal or state air-quality standards. (A few states do try to limit some metal emissions through permit guidelines.) Some of the metals are suspected carcinogens, but standards have not been established because data on their health effects are too scarce. Lead can be especially dangerous since the fine particles typically emitted from incinerators lodge in the lungs, where they are absorbed into the bloodstream and accumulate in bone marrow. High levels of lead are also routinely found in the ash residue from incinerators.

Japan has focused much effort on reducing emissions of hazardous metals, especially mercury, from incinerators. Batteries containing mercury and cadmium are no longer supposed to be burned, for example. The Japanese have also required battery makers to reduce the amount of mercury their products contain by five-sixths.

Scrubbers Are Crucial

Glass and metal are not the only substances that cause problems during combustion. Plastics, especially those that contain PVC, can create dioxin and toxic acid gases such as hydrogen chloride when burned. In fact, a report prepared within the New York State Department of Environmental Conservation concludes that U.S. refuse-burning facilities may emit more than 40 times the hydrogen chloride generated by coal-burning facilities. This highly corrosive gas irritates the eyes and respiratory system, contributes to acid rain, and corrodes metal and air-cleaning components in the plant itself. Unfortunately, the U.S. Food and Drug Administration is now proposing to allow more PVC in food packaging, ignoring the problems associated with disposing of this material.

Monitoring devices at three of the eight Japanese incinerators I visited indicated that levels of hydrogen chloride were "non-detectable," and all these plants removed 95 percent of this substance from the flue gas. The Japanese achieve this by presorting some plastics and employing scrubbers that spray a lime-and-water mixture into the flue gas. The acid gases mix with the lime slurry to form non-acidic, non-toxic calcium salts. Such a system also controls sulfur dioxide, another contributor to acid rain. Monitors at the Japanese plants indicated SO_2 emission levels of 2 parts per million (ppm), a removal efficiency of more than 95 percent. Equally important, all Japanese facilities have on-site treatment plants—powered by energy from the incinerator—to neutralize any contaminated water generated by the scrubbers or other plant components.

Of the 70 high-volume incinerators operating in the United States, only 2—in Framingham, Mass., and Marion County, Ore.—employ a similar scrubbing procedure. The latter plant has reduced hydrogen-chloride emissions to the extraordinarily low level of 10 ppm. In contrast, the incinerator in Peekskill, N.Y., emits 500 to 600 ppm of hydrogen chloride, a level typical of scrubberless plants throughout the U.S. As more trash-burning plants come into use, total hydrogen-chloride emissions will increase substantially unless scrubbers are installed.

Scrubbers have an added benefit: because they cool the flue gases, they allow any metals to adsorb onto the submicron particles—called fly ash—that are the residue of the burning process. This is important because recycling cannot separate all metals from the waste before it is burned.

Special efforts must be made to control mercury, since it will adhere to the fly ash only at temperatures below 140° C (284° F). Flue gas cooled to that extent will not rise as much from the incinerator and thus will increase the localized effects of whatever pollutants are emitted. Japanese incinerator operators have solved this problem by cooling the flue gas to remove mercury and then reheating it to make it disperse better.

A fabric-filter system (known as a baghouse) or an electrostatic precipitator (ESP) is used to collect the particles that have absorbed metals and other pollutants from the flue gas. The precipitator electrically charges the particles and captures them on a screen of opposite charge.

Dealing with Dioxin

No issue surrounding municipal incinerators is as controversial as emissions of dioxins and furans, and scrubbers can help here, too. The Hempstead, Long Island, incinerator was closed in 1982 partly because of dioxin emissions, and Denmark shut eight of its older plants in 1985 because of fears about the chemicals. These artificial organic compounds (specifically, polychlorinated dibenzo-dioxins, or PCDD, and polychlorinated dibenzo-furans, or PCDF) are suspected of causing a wide range of illnesses, from cancer to birth defects. Yet these effects are still subject to dispute, and the United States has not established any standards for incinerator emissions or human exposure. No method for testing dioxin emissions from incinerators is universally accepted, and even the question of whether to sample the flue gas or the fly ash is still being debated.

Despite this uncertainty, Sweden and other European countries have established strict limits on dioxin emissions. Sweden assumes that an acceptable daily limit is 1 to 5 trillionths of a gram per kilogram (2.2 pounds) of body weight. Swedish tests of mothers' breast milk a few years ago showed that babies were ingesting 50 to 200 times their daily limit. Intense concern over dioxin emissions from 27 resource-recovery plants, which burn half the nation's garbage, led the Swedes to establish a strict goal of 1 nanogram per cubic meter—soon to be mandatory—based on the performances of the best plants.

The only U.S. incinerator that comes close to achieving such results is Oregon's Marion County plant, which employs a scrubber and a baghouse. This plant registers dioxin emissions of .155 nanogram per cubic meter. A state-owned incinerator in Albany, N.Y., that does not use a scrubber or baghouse has registered 16 nanograms per cubic meter—over 100 times the amount emitted by the Marion County plant.

Installing scrubbers and high-efficiency particulate-removal systems can add $5 to $10 to the $20 to $35 tipping fees that plant operators charge for burning each ton of garbage. Many in the resource-recovery industry maintain that such costs are too high. However, these technologies are crucial if hazardous emissions are to be controlled and community concern about municipal incinerators addressed. Objections to installing scrubbers are particularly misguided since they may pay for themselves in the long run. The devices can be connected to heat pumps to capture 15 to 30 percent more heat from the flue gas, which can then be used to create more steam and electricity.

Disposing of Toxic Ash

Once the fly ash is collected by the baghouse or ESP, it must be disposed in secure landfills so the dioxin and heavy metals cannot contaminate water supplies. (Bottom ash—about 80 percent of the plant's total residue—does not pose as much of a problem. With good combustion, it is inorganic and not highly toxic, although small amounts of dioxin can be present.) Dioxins and furans adsorb so strongly to fly ash that acid rain and snow do not cause leaching, but they could affect the metals in the ash. Household liquids, including ammonia and turpentine, could also cause pollutants, including dioxins, to leach from the ash.

Little research has been done to determine the actual results of mixing household waste with incinerator ash, but Swedish and Japanese officials have established a prudent policy of keeping the two separate. The Swedes often place their ash in 55-gallon drums before sending it to landfills, while over 90 percent of Japanese landfills do not accept both kinds of waste. The Japanese frequently mix their fly ash with cement and deposit the blocks or pellets in landfills with liners and wastewater-treatment facilities.

The U.S. Environmental Protection Agency does not regulate the disposal of incinerator ash because it is still trying to figure out how to test its toxicity. Federal law now exempts municipal waste from being classified as hazardous, and the waste-to-energy industry is arguing that this exemption also applies to incinerator ash. This claim is highly controversial, especially since test results in New York State show that much incinerator ash has hazardous characteristics. Most states seem to be waiting for the EPA to act, with the result that handling practices vary from plant to plant. The most common practice is to mix fly ash and bottom ash and send them to landfills that also accept unprocessed garbage. The assumption is that the bottom ash will act as an alkaline buffer to prevent dioxins and metals from leaching. However, studies designed to determine if this really works and the best combination of bottom and fly ash are widely disputed and often flawed. And few attempts are made to prevent household chemicals from combining with the ash. This policy is extremely shortsighted and further guarantees citizen opposition to garbage-burning plants.

Worker Training and Plant Monitoring Essential

Whether an incinerator achieves low levels of dangerous emissions depends not only on its technology and presorting of the wastes, but also on how well workers are trained and the plant is monitored. Some U.S. incinerator operators try to hire workers with "steam experience"—those who have operated coal- or oil-fueled power plants. This move is based on the erroneous assumption that a municipal incinerator is primarily involved in producing energy rather than disposing of waste. Japanese incinerator workers inevitably describe their main job as preventing pollution, while U.S. workers see their role in terms of providing energy. One shift supervisor at the Delaware plant said, "It's not garbage, it's solid fuel." The high emissions levels of most U.S. incinerators attest that this attitude is prevalent.

Burning garbage, which is often wet and composed of many different substances, is more problematic than burning homogeneous fuels such as oil and coal. Workers need special training in controlling emissions and in protecting themselves from contact with harmful bacteria and toxic fly ash.

Employees at resource-recovery plants in West Germany, Japan, and Switzerland are thoroughly trained. Japanese workers spend 6 to 18 months learning how toxic chemicals are stabilized in the furnace and captured in the stack, and they must have an engineering degree and undergo on-site training. Similarly, German incinerator workers attend a training school operated by the Boiler Manufacturers Association. This training includes two years of practical experience and 6 months of theoretical work on combustion efficiency and dioxin formation and reduction. The

United States, in contrast, maintains no institute specifically for training workers to burn heterogeneous fuel, although the American Society of Mechanical Engineers is now working to establish such a program.

Monitoring and regulation of incinerators is most comprehensive in West Germany. Regulated emissions include acid gases such as hydrogen chloride and hydrogen fluoride, other contributors to acid rain such as sulfur dioxide and nitrogen oxides, and toxic metals, including cadmium, mercury, lead, chromium, copper, and nickel. Particulates and carbon monoxide are also regulated. Operators install many different instruments to monitor these emissions. In addition, all 46 West German resource-recovery plants will soon be connected by computer to state environmental agencies. Those that exceed standards can be and have been closed, and operators who knowingly violate regulations receive mandatory jail terms. Japan has established a comparable system.

Americans have much to learn from their overseas counterparts about handling solid waste without undue risk to human health. U.S. municipal waste disposal has been one of this century's bargains, but we are beginning to pay the price for past negligence. The technology exists to make waste-to-energy plants an effective way to reduce the volume of material dumped in landfills while limiting hazardous emissions. The next step in protecting the public's health must be to establish an integrated system of solid-waste management keyed by strong regulations that are swiftly enforced.

RECYCLING: COMING OF AGE[5]

The bowl of evergreens encircling a short stretch of Route 101 in southern New Hampshire might seem an unlikely spot for a garbage crisis. The ample open land suggests that somewhere among the slight hills could be found a landfill site where trash could be unobtrusively sequestered.

[5]Reprint of an article by Barbara Goldoftas, writer on science and technology. Reprinted by permission of the author from *Technology Review*, 90:28–35+. N./D. '87. Copyright © 1987 by Barbara Goldoftas.

And Wilton, N.H., might seem an unlikely town to have chosen mandatory recycling to solve its solid-waste dilemma. Residents of the area are fairly conservative, many of them longtimers joined by an influx of people largely employed by the region's high-tech companies. People living there seem to take the state motto "live free or die" entirely to heart. They don't like to be told what to do and resist government involvement in their lives.

Until 1979, Wilton did depend on a landfill of sorts—an old stone quarry where the town had long dumped its trash. As in many rural communities, the landfill was "treated with benign neglect," says town councillor Greg Bohosiewicz. Fires were regarded as "acts of God" that lowered the ever-growing piles of garbage, he points out. The strategy worked until the mid-1970s, when the state began pressuring Wilton to close the landfill because it sat on the banks of a river. The town faced a number of options, none of them easy: open a proper landfill, export the waste, incinerate it, or recycle it. Enter Bohosiewicz, then "just an interested citizen."

"The only thing I knew about trash at the time was that it magically disappeared someplace," he says. As a citizen, he wanted to keep taxes low. As an economist by training, he quickly realized that recycling was the cheapest alternative for dealing with the town's trash. Working as a "one-man committee," he sold the idea of a joint recycling center to six towns in the area. The center opened in Wilton in 1979, accepting materials ranging from aluminum foil to bottles and cans, plastic milk jugs to compost. Today about 65 percent of the area's population recycles—a high participation rate for a drop-off program. Wilton spends about $36 per ton to dispose of its waste, compared with $120 per ton in a neighboring town that does not recycle.

The Wilton program is part of a growing national trend toward recycling. At least 14 states, primarily in the Northeast, West Coast, and mid-Atlantic regions, have passed legislation that either promotes or requires recycling of residential garbage. More than 500 communities offer curbside collection of glass, paper, metal, and other materials.

Oregon recycles about 20 percent of its waste, and all towns there with 4,000 or more people must offer curbside collection of recyclables. New Jersey, which recycles about 15 percent of its rubbish, passed a law this spring requiring towns to set up recycling programs for leaves and at least three "marketable waste materials" of their choice. New York State and Philadelphia recently set goals to reduce their garbage by 50 percent by 1997, partly through recycling programs.

Recycling appears to be entering a new era both politically and economically. The wave of programs that accompanied the environmental movement in the 1970s tended to be small, private ventures that relied on volunteer labor. Many struggled because revenues proved difficult to make from limited quantities of used glass, paper, metal, and the like.

Today profits are generally not the bottom line, and environmental concerns are not the primary motivation. The new programs tend to be run by the public sector, which is spurred by steeply rising tipping fees—the cost of unloading garbage at landfills or incinerators—and an overabundance of garbage that has no place to go. The private sector, which is active in processing, incineration, and—in "bottle-bill" states—collecting bottles and cans, has steered clear of municipal collection programs, mainly because lower disposal costs generally do not accrue to private companies.

"Wherever there aren't huge open spaces, waste disposal costs are the fastest rising item on any municipal budget," says John Schall, the recycling director at the Division of Solid Waste in Massachusetts. "In the Northeast it's starting to cost so much for disposal that you don't even need to bring in revenues for the recyclables," says John Purves, former director of solid-waste disposal for New Jersey's Camden County. "You just need to get them out of the waste stream."

Many officials facing solid-waste crises see recycling as a way to dispose of garbage without raising taxes. Judy Roumpf, publisher of *Resource Recycling*, one of two magazines devoted to the topic, says that the public sector has come to the "gradual realization that although recycling costs, it is less costly than collection or disposal of waste."

Advent of the Throwaway Culture

Public recycling programs have been in place for years in countries such as Germany and Japan—in part because they are strapped for both raw materials and space. Communities in both these countries recycle 50 to 60 percent of their solid waste.

Bernd Franke, head of the Institute for Energy and Environmental Research in Takoma Park, Md., believes that public attitudes and education are critical to successful levels of recycling. Before recycling was introduced to several German towns, he says, "people asked 'Who is going to do this? They're too lazy.'" Now 95 percent of the population there recycles.

Some Tokyo suburbs recycle more than 50 percent of their refuse, with residents separating waste into numerous categories. In fact, recyclables there are not perceived as garbage. In much the way that Americans often sell or give away old cars, clothes, and books, Japanese citizens set aside materials such as glass, paper, metals, and household hazardous waste for processing and reuse.

Even in the United States, often criticized for being a throw-away, consumerist culture, indiscriminate disposal has a short history. In New York City in the late 1800s, scavengers were paid to "trim" the garbage of materials they could sell. In the mid-1890s, street cleaning commissioner George A. Waring tried to recapture some of these resources for the city to help defray the costs of waste disposal.

Called "the Apostle of Cleanliness," Waring understood the link between sanitation and public health and directed street sweepers to wear white uniforms to help the public make the same connection. Although incineration (then called cremation) was becoming popular, Waring introduced an ambitious recycling program, and in 1898 established the city's first plant to sort garbage and reclaim recyclables. According to a recent report by the New York State Legislative Commission on Solid Waste Management, between 1902 and 1924, 50 to 83 percent of U.S. cities separated some items for reuse.

During World War II, the loss of overseas sources of raw materials gave recycling a boost. Households routinely saved tin cans, glass, and other products. At least one-third of all paper was recycled during the war, along with copper, aluminum, and other strategic metals. Glass bottles and jars were refilled as many as 40 times.

As international trade in strategic metals resumed after the war, recycling declined. In the 1950s and 1960s urban "source-separation" programs were cut back, and by the mid-1960s most cities collected mixed garbage. Although some cities still picked up yard waste, separation of other materials was considered too expensive compared with the seemingly low cost of using landfills.

In the early 1970s, recycling began to rebound, spurred by concerns about diminishing natural resources, especially fossil fuels. Neil Seldman of the Institute for Local Self-Reliance in Washington estimates that more than 3,000 recycling centers were created during that time. Largely because of their size, these programs usually couldn't offer the reliable supply of materials necessary to attract secure markets. Most didn't survive the 1974–75 recession and the dip in prices for "secondary" materials that followed.

Wooing the Public

To ensure a steady supply of materials and reduce the volume of garbage, today's recycling programs try to enlist the active support of a large portion of the population. Approaches to wooing the public vary. In rural areas that lack the luxury of municipal refuse collection, residents already drive to the town dump or hire a hauler. So drop-off centers like Wilton's—where citizens bring all their refuse—require only the extra step of separating the recyclables. Center director Patricia Moore says that most people don't seem to mind the extra work, but those who do— about 35 percent—pay to have their trash taken elsewhere.

Each town pays a fee for using the center based on the size of its population, and also shares in the revenues from selling recycled materials. Moore estimates that Wilton saved $100,000 in disposal costs last year.

One of the center's outstanding traits is its cleanliness. Once a typical town dump, it now looks more like a summer camp with a tidy, litter-free parking lot, wind chimes, and a cluster of low, pre-fab barns. There's hardly an unpleasant odor. Wooden trays, each larger than a double bed, brim with aluminum cans, tin cans, brown glass, clear glass, and green glass. Hand-painted signs specify what goes where.

Bins at the front of one barn collect cardboard, newspaper, and mixed paper. At the far end of the lot stand a discarded vacuum cleaner and other appliances, a shelf bearing old books and a few shoes, jugs of used motor oil, several rows of car batteries, scrap metal, and a dumpster filled with organic waste for composting. People also drop off unwanted clothes, styrofoam packing pellets, and milk jugs and soft-drink containers made of PET (polyethylene terephthalate—one of the few recyclable plastics). A huge dumpster under a barn roof contains items that will be incinerated on site—disposable diapers, computer printouts, most plastics, and paper contaminated with food.

The center recycles about half the garbage it receives—a huge proportion for most such programs. About 35 percent is burned, and the remaining 15 percent, including the incinerator ash, is landfilled.

Urban Recycling Mecca

In contrast to the Wilton program, which collects many different materials, urban programs tend to focus on convincing a broad range of citizens to separate a few recyclables. Many programs offer curbside pickup during regular garbage collection and use extensive public education.

In the Northeast, New Jersey is almost a mecca for recycling. The most densely populated state, it faced landfill and toxic-waste problems early. It also contains an ample network for recyclables: dozens of industries, including paper and glass mills, use secondary materials, and more than 90 private dealers process waste paper, glass, and steel.

The state has increasingly pressured its 21 counties, which handle their own trash, to curtail the amount of solid waste they generate. In 1980 the state set a goal of reducing its waste by 25 percent and created a system of grants and loans to help counties launch recycling programs. The recycling office in Trenton, the state capital, also introduced Mr. R. E. Cycle, a magician featured on posters and bumper stickers who visits schools to teach students about the feasible magic of recycling.

Recycling has been mandatory in Camden County, N.J., since 1985. According to John Purves, former director of the county's program, "Before Camden, people didn't think that major recycling could work and be economically feasible." With a population of half a million, the county stretches from the relatively unpopulated Pine Barrens to the economically depressed city of Camden, which sits on the Delaware River opposite Philadelphia (whose own garbage problems often spill over into New Jersey).

Camden's recycling efforts were spurred in 1984 when a landfill used by 26 of its 37 towns threatened to close. Officials soon discovered that other nearby sites charged dumping fees three times as high. The solution: since 1985 all towns have had to recycle newspapers, aluminum cans, scrap metal, used oil, and yard waste. Some form of curbside collection is also mandatory. All but two towns pick up newspapers, and most collect bottles and cans, which they send to a central processing facility in Camden city.

Each town takes a different approach to encouraging citizen participation. Voorhees Township on the edge of the Philadelphia suburbs boasts a 100 percent participation rate. The tactic is simple: to have trash picked up, residents must put out the recycling pail distributed by the town, whether or not it contains bottles and cans. If they don't, the trash is left behind, labeled with a red tag.

The program in suburban Haddonfield has evolved into one of the county's most successful. The town opened a drop-off center for recyclables in 1981, and quickly moved to twice-monthly and then weekly pickups after dumping costs more than tripled. In addition to collecting bottles, cans, and newspapers, Haddonfield picks up "white goods"—old refrigerators, stoves, and other major appliances. Last year it collected 33 tons that were sent to scrap dealers in Camden city. The town also collects leaves and wood—2,200 tons and 1,000 tons, respectively, in 1986—chipping the branches, tree trunks, and other scraps for composting.

Each time curbside collection added another item or more frequent pickup, participation jumped. The drivers of the recycling trucks, who measure compliance with hand counters, estimate that 70 percent of all households recycle on any given day, and that only about 5 percent never do. Says borough administrator Richard Schwab, the drivers "know the town like the milkman"—who doesn't recycle, who recycles sometimes, which senior citizens share recycling buckets with their neighbors because they don't produce much trash. The town of 2,000 last year recycled 40 percent of its waste, saving $65,000 in tipping fees and reducing the total cost of waste disposal by $23,400.

Some communities in other states tackle recycling with a combination of programs rather than one comprehensive effort. A successful buy-back center in the Bronx pays people for their materials—an important incentive for many low-income residents. It accepts any quantity of most materials other than clothes and tires, including wood and several plastics. More than a year old, the program brings in over 100 tons per month.

In San Francisco, public and private programs each reach a distinct community or involve a different commodity. The programs almost sound like an array of rock groups: Encore picks up and sells wine bottles; Many Happy Returns buys back materials; Zoo Doo recycles zoo manure as compost. Others target Latin, black, and gay and lesbian communities. Recycling coordinator Amy Perlmutter says that the city's motto is "something for everyone." Her goal is to "figure out what would motivate a particular community and try to tailor solutions to suit the problems."

In Massachusetts recycling has been designed to work on a large scale. The Division of Solid-Waste planned a streamlined program, now operating in several communities, that ultimately is expected to be used across the state. Residents set out paper, glass, metal, and cardboard in a single container. The communities must make recycling mandatory, collect the materials, and offer ongoing public education. Regional processing plants similar to the one operating in Camden will separate and market the materials. "We created a model that is generalizable," says recycling director Schall. "Each community won't have to reinvent the wheel."

Garbage In, Resources Out

Early recyclers learned the hard way that thousands of households won't necessarily provide the clean, uniform product that manufacturers can use. Some people will throw away half-full jars of mayonnaise and marmalade, while others will scrupulously wash and dry their bottles and cans, removing the labels and squashing the tin cans flat. "A glass mill can't decide to use cullet [ground glass] over raw materials if it doesn't know what the quality of the glass will be, whether or not it will be contaminated by ceramic material or stones," says Schall.

The Wilton center uses a labor-intensive approach to ensure the quality of its recyclables: workers comb through the collection bins before pushing the materials into storage bunkers. Director Moore says that the staff learned not to overwhelm people with attention. "At first there was one attendant per category. But some people felt badgered. The attendants were idealistic and wanted everything right. You have to be careful when you're changing people's long-time habits—it's not done easily."

Workers at the center compress the cans and paper into tight waist-high blocks with a vintage 1921 chain-driven baler. Glass and plastic are stockpiled. The glass is first crushed by a tractor to reduce its volume, and the PET bottles and jugs are fed into a granulator that chews them into fingernail-sized bits. Most of these materials are used by industries in the region.

A higher-tech approach—but one that is still fairly simple—is key to the success of recycling in Camden County. In a matter of minutes, machinery at the Camden Recycling Facility sorts, crushes, cleans, and flattens recyclables into a form that manufacturers can use. The technology was developed by Peter Karter, formerly a civil engineer in the nuclear industry, who guards the details of his proprietary process carefully.

The facility is located on the site of an old scrap dealer. Outside the main building sit several neat 10-foot piles of color-coded cullet and crumpled aluminum. Trucks deliver loads of glass and metal throughout the day.

Inside the building workers wear ear plugs for protection against the din. Unsorted glass and metal travel up a long conveyor as a worker picks out toasters, wire hangers, pots, aerosol cans, paper bags, and other materials that don't belong. At the top of the belt a magnetic separator pulls out tin-plated steel cans and sends them to a processor that cleans and flattens them. The lightweight aluminum cans are blown into another processor and crushed into tight balls. The glass continues along the conveyor past workers who separate it by color—clear, brown, green. The bottles fall through vertical crushers into rotating sieves called trommels, which sift out labels and errant remnants of food. Along with other remaining garbage—about 10 to 15 percent of the original load—this refuse is landfilled.

Karter calls the recyclables raw materials that he mines. He sells the ground cullet to glass manufacturers in New Jersey and Massachusetts. Reynolds Aluminum, which provided the crusher for aluminum cans, buys the tight wads the machinery ejects for a plant in Hartford, Conn. The tin cans go to a plant in Baltimore that removes the tin plate from its steel base. Although tin cans are not recycled nearly as actively as aluminum, they still draw customers. Cans provide the only domestic source of tin in the United States; the steel also is valuable because of its high quality and consistent composition.

When the Camden Recycling Facility first opened in April 1986, only about 6 towns brought their recyclables there. By last March, when it reached its expected level of 40 tons per day (half its capacity), the number of towns had also climbed to about 40. The towns use the facility for free. If they set up efficient collection systems that cost less than the tipping fees they avoid, they can end up paying less for garbage overall. As the volume produced by the plant builds, the towns will also share profits from its sales.

Processing plants are a key part of the new wave of recycling. Massachusetts, New York, and several other states are planning to build regional facilities. By readying secondary materials for manufacturing, such technology makes large-scale recycling possible. The central facility provides a direct link between recycling in the home and recycling at the factory, fostering participation at the one end and demand for materials at the other.

To Market, To Market

One of the strengths of today's recycling enterprises is their intent to forge long-term links with industry. The Wilton program originally had trouble finding markets because of its size, and sometimes had to stockpile paper and glass. Now the New Hampshire Resource Recovery Association, a private nonprofit group that assists recycling efforts, coordinates the sale of most materials. Similarly, central processing facilities often do their own marketing, removing that burden from the public sector and individual programs.

If offered a steady supply of good products, industry can use recycled and processed materials in place of their virgin counterparts, and often advantageously. Cullet can replace some or all of the sand, soda ash, and limestone a mill ordinarily uses to make glass. Before energy prices shot up in 1973, manufacturers generally used just 15 to 20 percent cullet. Some mills will now use 80 to 90 percent if it is clean and free of contaminants. Because it melts at lower temperatures than the raw materials, cullet saves wear and tear on the melting tank and uses less energy. It also lowers emissions from the furnaces.

The most active advocates of aluminum recycling are its manufacturers: they use 95 percent less energy to recycle aluminum than to smelt it from bauxite ore. Reynolds and Alcoa promote recycling throughout the country and support existing bottle bills. Last year Alcoa alone recycled 550 million pounds of aluminum, and more than half of the 300 billion aluminum beverage cans purchased since 1981 have been reclaimed. According to the Worldwatch Institute, on average a recycled aluminum can is back on a supermarket shelf in six weeks.

Unlike glass and aluminum, paper cannot be recycled indefinitely—eventually the fibers break down. But many grades of paper can be de-inked, cleaned, and bleached, processes that require less energy and water than those used to create virgin pulp. The recycled material is used to make products such as game-boards, cereal boxes, covers for hardcover books, ticket stubs, and tissue paper.

There is a sizable demand for waste paper. About 200 of the approximately 600 U.S. paper mills use waste paper exclusively, and another 300 use 10 to 30 percent. Rod Edwards of the American Paper Institute claims that the domestic use of waste paper is expanding. Countries in Europe, Asia, and South America with little timber are also importing growing amounts of waste paper, one of the largest exports from New York Harbor.

Recycled paper still sometimes carries a stigma, says Jeff Coyne of Earthworm, a nonprofit organization in Boston that has recycled office paper since 1970. One of Earthworm's customers uses recycled paper for all its products but one—surgical gowns. The customer worries that gowns made from waste paper might be seen as less than sterile.

Securing markets for secondary materials wins just part of the battle, because the prices they command fluctuate with changes in foreign demand and the overall state of the U.S. economy. A growing economy boosts demand and thus prices. Construction contractors, for example, begin to need more plasterboard, which is faced with waste paper. Consumers also buy more pizza in boxes, more shoes in boxes, more stationery, more books. But when the economy drops into a recession and exports fall, the price also falls. Recyclers must be able to survive such roller-coaster prices.

Opinions vary on whether an increase in recycling and processing facilities will flood the markets, driving down the price for secondary materials. In New York, which produces 25,000 tons of garbage a day, public officials worry about just such a problem. But Massachusetts' Schall insists that markets for reclaimed glass, steel, and aluminum "will expand if you can assure manufacturers of a large, constant supply of high-quality materials. In cases where secondary materials compete directly with their virgin counterparts, the steady supply of a non-contaminated product will create its own demand."

Government subsidies on the use of virgin resources have often undercut the market for secondary materials. For example, tax credits generally have favored exploitation of natural resources like timber, sand, and bauxite. Tax breaks on freight costs and capital investments rewarded industry for using raw rather than recycled materials. New York, North Carolina, and other states have tried to remedy this situation by giving tax breaks to companies that use secondary materials or purchase recycling equipment. Florida and Wisconsin also offer sales-tax exemptions for firms that use recycled materials. Other states are considering ways to reduce the amount of garbage they produce—by curtailing the amount of packaging on products sold or manufactured within their borders, for example.

Some states are spurring demand for secondary materials by increasing their use by the public sector. New Jersey's recent solid-waste bill requires state agencies to use recycled paper "whenever the price is competitive." Maryland government buys half its bond paper and nearly all its tissues and towels from manufacturers that use recycled materials. Similar actions by the federal government influenced napkin producers to use waste paper. The voluminous *Federal Register* is also printed on recycled paper.

First Priority Rather Than Last

The future of recycling depends largely on the willingness of public officials to create new, long-term programs in the face of current pressures to build incinerators. Recycling is sometimes seen as a competitor of incineration because it can reduce the amount of waste that plant operators are paid to receive. Many communities guarantee incinerator operators a minimum volume of trash. But while recycling can eliminate high-quality fuel like paper and cardboard, it also removes unburnable glass and metal and reduces the amount of slag that forms from melted glass on the sides of furnaces. Slag and contaminated ash are such major problems that some incinerators have facilities on site that remove glass and metal from garbage before it is burned.

Yet some recycling advocates believe that attempts to see the two alternatives as complementary undermine the importance of reducing and reusing waste. "Too often recycling is seen as part of a happy marriage with incineration, rather than a primary solution," says Thomas Webster, research associate at the Center for the Biology of Natural Systems in New York. Webster has worked with biologist Barry Commoner on a pilot recycling program for East Hampton on Long Island designed to deal with about 70 percent of the trash—an amount comparable to incineration. Webster believes recycling could reduce trash even more if legislators set standards for packaging design, targeted plastic milk jugs and other items for deposit fees, and introduced "waste-initiator" taxes on packaging that can't be recycled.

"Reduction and recycling of waste should be first priority rather than last priority," says Webster. He considers incinerators a needlessly complex and expensive form of waste disposal—and a problematic one. The plants produce emissions believed to be harmful as well as toxic ash that must be landfilled.

"What we need to do is to get rid of the garbage, not just move it through one more machine," maintains Webster. "To get high levels of recycling you have to work hard. You need legislative strategies, and you need political will. But once you have the systems in place, you've solved the problem."

II. OCEAN WASTE DISPOSAL

EDITOR'S INTRODUCTION

Waste disposal is not an issue confined to the land, but extends to water as well, and particularly to the ocean, into which tons of industrial effluvia and sludge (treated human waste) have been dumped for years. A dramatic case is the New York Bight, a triangle of off-shore ocean between Long Island and New Jersey into which New York City has dumped its sewage—to the point where Long Island beaches are becoming insanitary.

In the first article in this section, reprinted from *Oceans*, Beverly Payton discusses the controversy over the dumping of sewage sludge in the New York Bight. As she points out, massive dumping of millions of tons of sludge, together with vast amounts of industrial waste, and millions of tons of raw sewage flowing into the ocean from the Hudson River have resulted in damage to marine life near the ocean floor. In a related article, Anthony Wolff, writing in *Audubon*, comments on the effects of wide-scale dumping in harbors and off-shore waters of many other American cities, and the demand of ecologists for an end to *all* ocean dumping. A following article by Tim Smart and Emily T. Smith, staffwriters for *Business Week*, focuses on the harmful effect on shellfish of coastal dumping. As they point out, however, the cost of cleaning up will be formidable—more than $75 billion just to overhaul waste-treatment plants, which produce 70% of all effluent discharged into coastal waters.

In "The Case for Ocean Waste Disposal," William Lahey and Michael Connor, marine research scientists writing in *Technology Review*, survey the recent history and legislation regarding ocean dumping, and its attractiveness as a much cheaper means of waste disposal than options available on land. They conclude that dumping in deep-water ocean is feasible, since dynamic currents of the sea in such areas quickly dilute, break up, and "clean" wastes; but they warn that more research and monitoring will need to be done to determine how much waste the ocean can assimilate. John W. Farrington, director of the Coastal Research Center at the Wood's Hole Oceanographic Institution, and three research colleagues, writing in *Oceanus*, draw similarly wary conclusions.

BLIGHT IN THE BIGHT: SEWAGE AND WATER DON'T MIX[1]

It is a crisp fall morning. Commercial fisherman Ed Maliszewski cruises the *Faye Joan* above an area called the Mud Hole. He trawls for whiting and flounder as they begin their southward migration to warmer water. After a five-mile tow, the trawl doors are pulled up and the net swung aboard. Besides the usual catch, the net contains a black, coarse, hairy substance that clogs the mesh. Ed has learned to put up with it. He makes his living in the New York Bight, an area some marine biologists have called the most polluted coastal waters in the world.

Last year more than thirteen million wet-tons of wastes were dumped here, making this section of the Atlantic Ocean between Montauk Point, Long Island, and Cape May, New Jersey, a pretty foul place. In 1983 it received 8.3 million tons of sewage sludge and 4.4 million tons of harbor dredge spoils. Allied Chemical dumped 38,000 tons of acid waste here. In addition, about 150 million gallons of raw sewage spews out of New York into the Hudson River every day and empties into the ocean.

All this turns the New York Bight into an organic soup seasoned with high concentrations of such toxic metals as silver, chromium, copper, nickel, lead, zinc, cadmium, and mercury. The area is also a hot spot for polychlorinated biphenyls—the famous carcinogenic PCBs.

According to researchers at the Northeast Fisheries Center in Sandy Hook, New Jersey, the bight's vital signs are low. As part of the National Oceanic and Atmospheric Administration's (NOAA) Ocean Pulse program, they have been monitoring the health of the continental shelf of the northwest Atlantic Ocean from Maine to Cape Hatteras, North Carolina. "The distribution

[1]Reprint of an article by Beverly M. Payton, free-lance writer specializing in science writing. Reprinted by permission from *Oceans*, 18:63–67. My./Je. '85. Copyright © 1985 by the author.

of contamination in the New York Bight is like a bull's-eye, with the greatest amount of contaminants close to the center of dumping activities," says John Pearce, chief of the Environmental Assessment Division at the Sandy Hook lab. "We've found stations that are nearly azoic, that is, there are almost no animals there. We've also found a lot of affected and stressed fish."

The stress ranges from genetic abnormalities in the Atlantic mackerel's eggs and larvae, to red blood cell mutations in windowpane flounder and red hake, to "black gill disease" in rock crabs and lobsters. In 1970, shellfish in a six-mile radius around the dump sites were found to have such high levels of bacteria that the Food and Drug Administration (FDA) closed the beds. The quarantine area was extended in 1974 and now encompasses the entire upper bight.

There have been legislative attempts to clean up some of the mess. Two measures, the Clean Water Act and the Ocean Dumping Act, made it through the House of Representatives last session but were never passed by the Senate. Hearings are now being held in both houses on new versions of the two bills. Meanwhile, the federal Environmental Protection Agency (EPA) announced on 1 April 1985 that it intended to close the sewage sludge dump site that is twelve miles offshore within the next eighteen months. This would force New York and New Jersey dumpers to barge their sludge to a site 106 miles away, beyond the edge of the continental shelf. New York City officials have decided not to challenge the closure in court, the threat of which, some environmentalists maintain, caused the EPA to postpone its decision for months.

"We're the local whipping boy on this issue," says Andrew McCarthy, spokesman for New York City's Department of Environmental Protection (DEP). "Our position is that we should remain at the twelve-mile site. It'll possibly cost up to $40 million more to take sludge out 106 miles, and that kind of money would be better spent on other environmental projects."

In an effort to keep the twelve-mile site, the city's DEP hired two firms, Ecological Analysts and SEAMOcean, to prepare a report on the effects of sludge dumping. The report states: "The impacts of sewage sludge dumping are minimal." It claims that environmental changes, now taken as indications of degradation, are all either natural occurrences or caused by contaminants from sources other than sewage sludge. "As a logical extension," says the DEP, "if sewage sludge dumping were halted at the twelve-mile site, there would be no significant improvement in either the environment of the New York Bight or the degree of protection from health risks afforded the adjacent metropolitan population."

The DEP has a point. The volume of pollution from raw sewage in the Hudson-Raritan River complex is considerable. Moreover, satellite images show that the river plume hugs the New Jersey shoreline, sweeping pathogens towards beaches. In its application to continue dumping at the twelve-mile site, New York City's DEP estimated that the Hudson-Raritan outflow contributes at least 500 times more sewage-associated microorganisms to the New York Bight than sewage sludge.

Pearce agrees that the riverine outflow probably poses as much of a public health risk as sludge dumping. But he believes the twelve-mile site should be phased out because of its impact on the benthic (bottom-dwelling) animals. "In a large area," he says, "about 50 to 100 square miles, the populations of bottom-dwelling fauna—that is crabs, lobsters, snails, clams and worms—have been altered because of sewage sludge dumping."

On a warm, still day in late summer, the *Kyma*, a research vessel for the Sandy Hook lab, casts off on a trip to retrieve forty-eight wooden boxes. Each contains a different sediment treatment that has been left for twelve days in forty feet of water off the coast of Long Island, well away from the sludge dump site. From his shirt pocket, fisheries biologist Clyde MacKenzie pulls out a crumpled paper which lists the unappetizing recipes: fresh sewage sludge, sterilized sewage sludge, organic peat, week-old sewage sludge, month-old sewage sludge, plain sand, plain fuller's earth (a white clay), fuller's earth laced with mercury, and more. The researchers aboard the *Kyma* are trying to find out which component of sewage sludge is the most repugnant to benthic animals from the variety and number of marine invertebrates that settle on each of the various treatments. "What we're trying to do," says Clyde, "is simulate the real world in a controlled environment."

Theoretically, sewage sludge does not accumulate on the ocean floor in the real world. Dispersed in the turbulent wake of the dumping barge, it is supposed to be suspended in the water column above the *pycnocline*, a density gradient caused by a differ-

ence in either temperature or salinity. Whatever manages to settle to the bottom, the theory says, should spread out over a wide
area. On the contrary, says John Pearce: "The oceans are big and
have lots of water to dilute things, but things don't get uniformly
diluted." This is especially true in the New York Bight.

The sludge dump site is situated at the edge of the Christiaen-
sen Basin, a natural accumulation area. Aboard the *Kyma*, fisheries chemist Andy Draxler says, "We've made a lot of grabs out
there, and the sediment is just a black, gelatinous ooze the consis-
tency of mayonnaise." From the basin, contaminants end up 110
miles down the Hudson Canyon. The lab collected sediment samples at stations all the way out to the edge of the continental shelf
slope and found unexpected high levels of PCBs and petroleum
hydrocarbons there.

The steady hum of the *Kyma*'s engines are cut to a grumbling
idle as she approaches a cluster of white buoys, each with the
NOAA logo. Andy dons his scuba gear and braids his chest-
length beard to keep it from floating in his face.

When researchers from the Sandy Hook lab dive in the dump
site area proper, they must take special precautions. Instead of
scuba, they use diving helmets and surface-supplied air. When
they come back aboard the *Kyma*, they are immediately hosed off
with fresh water and doused with a disinfectant. Then, dry suits
are peeled off and thrown to soak in a tub of bleach, and the divers wash again with a disinfectant. MacKenzie says he once ne-
glected some of the washing procedures after a dive in what was
thought to be an uncontaminated area. He became so sick that
he required hospitalization. "I was out of work for over a month,"
he says. "I thought I'd never be well again."

As Andy climbs down a ladder hung over the starboard rail,
Clyde hands him a wooden board with syringes mounted securely
under strips of plastic tubing. He will use these to sample the water just above four of the sediment boxes. In these, he attempted
to create a sulfide-generating layer by including sulfide paper and
a nutrient medium composed partially of exploded cellulose (to
simulate toilet paper).

When bottom sediments become anoxic (lacking in oxygen),
anaerobic bacteria take over the job of decomposition. Their res-
piration generates hydrogen sulfide, a gas that smells like rotten
eggs. Andy wants to find out if benthic animals will settle on sediments that reek of sulfide. Some recent studies suggest that cer-

tain organisms will not settle unless the sediment is generating at least a little sulfide, while other organisms, such as lobsters, do not seem to care one way or the other. But for most animals, sulfide is toxic. "A lot of sulfide is generated in the sludge area," Andy says. "It's enough to kill certain animals."

Not all species of marine life in the New York Bight suffer from the sewage. Phytoplankton love it. Fertilized by both the heavy nitrogen and phosphorus content of sewage and the up-welling of organic-rich currents from off the continental shelf, phytoplankton multiply so rapidly in the spring that the zoo-plankton cannot consume them fast enough. By summer, much of the phytoplankton dies and sinks to the bottom where bacteria, decomposing the organic overload, use up the available oxygen. "We've been monitoring the dissolved oxygen levels for more than a decade," says Pearce, "and almost every summer in the part of the New York Bight apex that receives the sewage sludge, the bottom-dissolved oxygen drops close to zero." Because of its lack of marine life, local fishermen have dubbed an area five miles square and fourteen miles east of Sea Bright, New Jersey, the "dead sea."

In 1976 many thought the "dead sea" had spread throughout the New York Bight, when a severe anoxic event decimated all the benthic organisms in an area of about 8,600 square kilometers along the New Jersey coast. At the same time, an unusually large volume of sewage debris washed up on Long Island beaches. Many erroneously blamed the fish kill on sewage sludge dumping.

In 1977, when the Marine Protection Research and Sanctuaries Act was up for reauthorization, the fish kill and fouled beaches were still fresh in legislators' memories. Congress passed an amendment to that act, forbidding ocean dumping after 31 December 1981. It almost worked. Camden, New Jersey, developed an effective composting system, and Philadelphia began shipping its sewage sludge to strip-mined areas for use as fertilizer. Current studies reveal that the closed Philadelphia dump site is now undergoing a slow, continuing recovery. But New York City took the EPA to federal court and won an indefinite extension because the EPA failed to prove that sewage sludge "unreasonably degraded the environment."

Part of the problem, according to Commander Carl Berman of the NOAA Commissioned Corps at the Sandy Hook lab, is that biologists have not developed a precise definition of what unreasonable environmental degradation is. "What we need," he says, "is a degradation index, so that researchers can take measurements into the courtroom and say, 'This should be 8.8 and it's a 2.1; our cutoff point is 4.0, therefore it's unreasonably degraded.'"

In September, at the Oceans '84 conference held in Washington, D.C., Paul Davis of SEAMOcean presented a global comparison of contamination in estuaries and coastal waters. He said pollution in the New York Bight is not out of line when compared with other coastal areas having a high population density and heavy industrial base. He added that the New York Bight is only of moderate concern when normalized to its population density. In counterattack, Berman argued: "What you're saying sounds like a doctor who says to his patient, 'Well, I've done a comparative study and I've found that this patient has cancer, and so does this one, and so does this one, so you can relax because your cancer isn't any worse than anyone else's.'"

During the summer, New York and New Jersey routinely sample the waters along their shorelines for the presence of fecal coliform bacteria. If New York gets a count of more than 200 per 100 milliliters of water, the beach is closed to bathing. New Jersey's standard is even tougher, requiring that the fecal coliform count not exceed fifty. New York City's DEP cites the relatively rare instances of beach closure as evidence that the water quality is acceptable.

But many biologists believe that fecal coliforms are not a good indicator of the presence of human pathogens. The reason is that most bacteria die within a few hours after exposure to salt water, but many viruses do not. Further, most species of *Acanthamoeba* form resistant cysts and have been found to survive almost three years after refrigeration. These pathogenic protozoa feed on bacteria and are commonly found in sewage-impacted environments. They can cause severe illness in healthy people and may even be fatal to those having an impaired immune system.

Other studies have shown that large numbers of *Bacillus* bacteria from near the dump sites of the inner New York Bight are resistant to mercury and the antibiotic ampicillin in concentrations that would kill normal *Bacillus* strains. In addition, some studies done at the University of Maryland suggest that certain species of bacteria and virus taken from the ocean may remain viable in a dormant state but go undetected because the cells are unable to grow on conventional culture medium.

With the day's fishing lost, the *Faye Joan* turns back toward Belford, New Jersey. Along the way, the fouled net is reversed and dragged on the surface to remove as much of the fibrous gunk as possible. Too often, that is not enough. The men have to haul it ashore, spread it on a concrete apron, and scrub it down with wire brushes. Nobody breathes deeply: the net smells like rotten eggs.

"We're the endangered species here," says Ed Maliszewski, echoing the concerns of many other commercial fishermen. Fishermen also complain that in the past few decades the quantity of fish in the bight has dwindled. New York City's DEP accuses them of overfishing. Yet, studies found striped bass, bluefish, white perch and eels all with high PCB levels. Both New York and New Jersey now issue consumption advisories to recreational fishermen. Following these announcements, charter boats are emptier, commercial landings fewer. Says Ed, "If something isn't done about the dumping soon, fishermen will be the first to go."

Fishermen are not the only ones who suffer the economic impact of marine pollution. "The tourist industry in New Jersey is very lucrative," says Derry W. Bennett, executive director of the American Littoral Society, a national marine conservation group headquartered at Sandy Hook. "If there is even a perceived threat that the waters are not safe for swimming, people here lose money." He adds that bathers frequently encounter a greenish, stringy slime that wraps around their arms and legs, clings to bathing suits, and smells. "The locals have a very graphic name for it—'sea snot'. Actually it's rotting plankton, usually long-chain diatoms." Bennett believes the overabundant plankton is another result of organic pollution. "If we cleaned up our act we wouldn't get this as often," he says. "The system's out of whack when you get huge quantities of these organisms; it's a sure sign of a biological system that's not working well—high numbers, low diversity."

All of the proposed solutions to the problem of ocean dumping seem to come with their own set of problems. Environmental groups, such as the American Littoral Society and Clean Ocean Action, would prefer to see land application of all sewage sludge. But when the New York City DEP studied this option in 1981, it concluded that this alternative was prohibitively expensive. Also, it carried its own human health risks, such as airborne pathogens and the leaching of metals and other toxins into groundwaters. A similar problem exists with land-based incineration, posing the conventional threat of air pollution and the dilemma of what to do with the ash residue.

At present, it seems that barging sewage sludge beyond the continental shelf will be the option chosen. Pearce thinks this is an acceptable alternative. The water there is more than 6,000 feet deep, so the wastes will be suspended in the water column longer and dispersed better. Also, the deep ocean is not as densely populated with marine life, Pearce says. "The contaminants will be less likely to harm animals involved in food chains that culminate in man."

But not all biologists believe the deep ocean is a better place for waste. Dr. Howard Sanders of Woods Hole Oceanographic Institution points out that marine life in the deep sea, having evolved in a static environment, have a much narrower range of tolerances than life found in shallow coastal waters. He says, "The organisms that evolved in the deep sea can stand only a very small temperature, pressure, salinity, or oxygen change. In high-stress coastal areas, organisms have evolved that can respond quickly to the aftermath of a catastrophe. They grow rapidly, produce vast numbers of young, and saturate the environment. But in the deep sea everything is going at a much slower pace."

While gearing up for his dive to help Andy retrieve the sediment boxes, lab technician Jim Duggan peers at the ocean from the starboard of the *Kyma*. Barely visible on the horizon, a barge, heavy with sludge, rides low in the water. Jim hopes a dumper does not pass any closer because the revolting smell of decaying organics travels a long distance downwind. Also flies, following the dumpers out to sea, sometimes jump ship. "They just about eat you alive," he says.

Jim climbs down the ladder and disappears below the surface. From the *Kyma*, Andy maintains radio contact with him. Shortly, Andy is hauling a cable hand over hand as he lifts the last stack of sediment boxes from the ocean floor and gingerly sets it down among the cluster on the aft. Later, he and Clyde scrape the top layer into plastic jugs. Back at the lab, the contents will be screened for a census of marine invertebrates, which will show that these animals avoid any of the boxes containing sewage sludge.

They're not the only ones. On the *Kyma*, everyone but Clyde and Andy avoids the aft. It smells like rotten eggs.

Update: THE YEAR THE BIGHT DIED

Localized fishkills sometimes occur in the New York Bight during summers, but the 1976 episode is the worst on record. Between July and October of that year, almost all the benthic (bottom-dwelling) animals off the New Jersey coast died. Scuba divers found the lifeless bottom covered with a flimsy flocculus, resembling brown snow.

During the same period sewage debris washed up on Long Island beaches. It formed a five-foot swath running parallel to the shoreline. Black and brown balls were attached to much of the debris, which at first were thought to be raw human feces. Later examination showed they were composed of tar and grease with high concentrations of fecal coliform bacteria. On June 16 all state beaches were closed and a week later the governor of New York declared Nassau and Suffolk counties disaster areas.

The 1976 episode attracted wide public attention. Many believed humanity had finally saturated the ocean with its waste and now the ocean was giving some of it back.

Biologists now know that an unusually warm, calm spring, together with persistent southerly winds, encouraged a population explosion, or bloom, of the dinoflagellate *Ceratium tripos*. In July, these phytoplankton started to die and sink to the bottom, aggravating already anoxic conditions.

Because the bloom spread over such a wide geographic area, nutrient enrichment from sewage in the New York Bight is no longer thought to have been the cause. "If ocean dumping had a role in that event, it was in a very local sense," says John Pearce, head of the Environmental Processes Division at Woods Hole Oceanographic Institute in Massachusetts. "It may have intensified the already anoxic conditions in the Bight Apex."

But evidence of yearly decreases in bottom dissolved–oxygen, usually peaking in late summer, has caused some researchers to believe that the bight may experience other fishkills. They say the system may be changing in such a way that even a slight imbalance, caused by either natural events or waste loadings, could

drive it again toward anoxia. (By Beverly M. Payton. From
Environment, 27:29, N. '85. Copyright © 1985 by the author.)

FECAL FOLLIES[2]

To think about New York City's sewage sludge problem, you
start by thinking of the Big Apple as the Big Toilet. (No problem
there: That's how the rest of the United States thinks of the city
anyway.) Now think of the Big Toilet flushing. Think of all that
you-know-what being slightly refined into a terrible black goo:
That's sludge. Now think of all that sludge being dumped into the
briny waters of the New York Bight, just outside lower New York
Bay, practically under the nose of the Statue of Liberty. Now
you've got the sewage sludge problem.

Last April the Environmental Protection Agency, which has
jurisdiction over such matters, finally promulgated the final solution to the problem. Forthwith, sludge dumping is to be relocated
from its present site, only ten miles out from the nearest shore
in eighty-eight feet of water, to a site at the very edge of the continental shelf, 106 miles out and more than three thousand feet
deep. Having exhausted their opportunities to petition, appeal,
litigate, and procrastinate, New York City and the neighboring
sewage authorities that use the bight as a septic tank have reluc-
tantly acceded to EPA's decision.

The solution fails to satisfy anyone. Ecological purists are
dead sure that a civilization that dumps its detritus in the mater-
nal ocean, no matter how far out or how deep down, is hardly civilized. They insist on an end to *all* ocean dumping, the sooner the
better, before the natural system takes some as-yet-unspecified
revenge. The pragmatists, who include most of those who actually face the necessity of doing something with the sludge, tend to
the self-serving conviction that throwing the stuff in the bight
hasn't hurt anyone yet—at least not so as you can prove it—and
it's the cheapest way, so why change? The dispute has eaten up
ten or more years; the careers of scientists, bureaucrats, and poli-

[2]Reprint of an article by Anthony Wolff, *Audubon* staffwriter. Reprinted by permission from *Audubon*, 88:32–35. Ja. '86. Copyright © 1986 by *Audubon*.

cy-makers; tons of paper; and tens of millions of dollars—all to achieve a compromise whose outlines, at least, could probably have been discerned at the outset.

Dumping municipal sewage sludge in New York Bight is nothing new. Since 1924 the area designated for the purpose has been the so-called 12-Mile Site in the New York Bight Apex, a corner of the Atlantic coastline formed by the New Jersey shore to the west and Long Island to the north. The city, which began using the site in 1938, currently shares it with eight other sewage authorities from Long Island and New Jersey. Together they account for some 95 percent of all the municipal sewage sludge that is dumped in the ocean from the United States.

By 1973 some 4.6 million wet tons of sludge were being dumped in the bight annually. Ten years later, increased sewage treatment (avidly promoted by EPA) had almost doubled the sludge burden to 8.3 million wet tons. "I think things are getting better," says Dennis J. Suszkowski, chief of EPA's regional Marine and Wetlands Protection Branch, where the sludge problem came to rest. Suszkowski's master's thesis was entitled *Water Pollution in New York Harbor: An Historical Perspective*. "Dissolved oxygen [a standard measure of water quality] was worst in the 1930s," he says, "before the beginning of sewage treatment. Everything was raw sewage." Even now, Suszkowski adds, "the biggest impact, in my opinion, is the large amount of nutrients you get out there," most of it from sources other than sludge. "The production of phytoplankton in the New York Bight is greater than off Peru," where the natural upwelling of microorganisms normally supports one of the world's great fisheries.

No one took notable offense at the sludge dumping until 1971, during Earth Week, when the bight site was indicted with great fanfare as one of the world's Ten Worst Ecological Disasters. In 1972 the federal Marine Protection, Research, and Sanctuaries Act (a.k.a. the ocean dumping law) transferred regulation of all ocean dumping from the permissive Army Corps of Engineers to the young, vigorous Environmental Protection Agency. Even so, the problem of putting sludge in the bight was not immediately apparent. EPA studied the matter, held some hearings, and routinely issued interim permits in 1974 and 1975 for dumping as usual.

Then the you-know-what hit the fan. Actually it hit the beaches of Long Island, in the form of what was officially described as "large quantities of floatables." (When you're talking sewage, you tend to speak in euphemisms.) Long Islanders were vexed, especially by the suggestion that the gobs of black goo and other nasties that were washing up on their pristine beaches were only the advance wave of a much bigger invasion. According to this theory, a mountain of fecal ooze, the concentrated toilet flushings of fifty years, was migrating inexorably across the ocean floor from the bight toward the Long Island strand. (See "Here Come de Sludge" by Gary Soucie in the July 1974 issue of *Audubon*.)

"We don't mind being New York City's bedroom," quipped Alfonse M. D'Amato, Hempstead's presiding supervisor on his way to the United States Senate, "but we don't want to be the city's bathroom."

No matter that it was quickly established that the flotsam was not jetsam from the sludge boats. Soon everyone was suing everyone else. Hempstead sued the city for damages, demanding the city improve its sewage treatment. The National Wildlife Federation sued EPA and the Corps of Engineers to do away with dumping altogether. Without waiting for Congress to amend the ocean dumping law in 1977, outlawing the deep-sixing of "harmful sewage sludge," EPA in 1976 announced that the 12-Mile Site would be closed as of December 31, 1981, and advised New York City and the other dumpers to make other arrangements.

Easier said than done. A favorite scheme, back when organic gardening was the universally understood metaphor for the virtuous life, was to turn the infected sludge into healthy compost. However, "compost made from sludge," says George N. Lutzic of New York City's Department of Environmental Protection, "is deficient in nitrogen. You don't end up with a fertilizer. At best you have a soil amendment, a conditioner, like peat moss."

Good enough; but what to do with it? One popular idea was to spread the stuff on the city's much-abused parklands, which would presumably be grateful for the tonic. No good: "All the property under the jurisdiction of New York," says Lutzic, "could be covered with a blanket of sludge compost in only seven years."

An alternative was to ship trainloads of New York's ordure upstate or, as Philadelphia had done, to western Pennsylvania, where it would be just the thing to fill in landscape scars left by abandoned strip-mines. The trouble was that no one asked the outlanders; they turned out, according to Edward O. Wagner, the Department of Environmental Protection's man in charge of sludge, to be less than enthusiastic about having their scenery redecorated with other folk's slightly refined you-know-what. Burning the sludge, another Good Idea, inspired the same objection: No one wanted the incinerator to be in his neighborhood. "We can't site it" is Wagner's response to people who offer such suggestions for getting rid of his unpopular product on land.

With time running out on EPA's deadline, the dumpers, led by New York City, sued to block the federal order. The judge agreed that EPA had gone too far too fast: The agency would have to consider applications for the redesignation of the 12-Mile Site. Meanwhile, the nearshore site would remain open.

The city's initial application for redesignation of the site, filed on July 10, 1980, was rejected. The city filed an amended application just in time for Christmas 1983. In two volumes, the amended application's several hundred pages weighed several pounds. It was, said the city's Department of Environmental Protection, "structured to be responsive" to EPA's requirements; its "protocols" were "developed through interaction with EPA"; and, "in lieu of specific guidance from EPA," it included "an innovative multimedia analysis which compares the human health, environmental, economic, and public perception effects" of alternative sludge-disposal methods. Ergo, concluded the city: The 12-Mile Site was obviously safer, cheaper, and more convenient.

Nothing doing, replied EPA, which had already designated a Deepwater Municipal Sludge Dump Site, 106 miles out, and was hardly anxious to change its mind. The feds waited a mere four months before announcing a "tentative" decision to deny the city's laborious application. Three one-day public hearings, followed by ten months of bureaucratic silence, sufficed for EPA to make its decision final.

"In our opinion, impacts at 12-Mile Site are not increasing," Edward Wagner still insists. "EPA did not agree."

The city, which uses its own sludge vessels instead of the private barges that service other dumpers, is moving toward compliance with all deliberate speed. The four vessels in the New York sludge armada are neither big enough nor fast enough to go the new distance, and none is certified for open-ocean use. By next April 7th, one year from the EPA order, says Wagner, one of his two 100,000-cubic-foot sludge vessels will be refitted for deep-sea service, taking about ten percent of the city's sludge to the 106-mile site. Its sister ship and the two 65,000-cubic-footers will continue hauling the rest of the muck to the 12-Mile Site until November 20, 1987. By that date, Wagner promises, New York will have three huge new ocean-going barges in service. All dumping at the 12-Mile Site will cease.

New York may have lost a costly and trying battle, but in a sense it won the war: Ed Wagner likes 106 miles better than no ocean dumping at all, considering the alternatives. "No one is saying that putting it in the ocean is *beneficial*," he allows, "but the question is, 'Is what we're doing environmentally sound?' If the answer is yes, there's no reason to change. It is our firm conclusion that, for New York City, the most environmentally safe method is disposal in the ocean. If ocean dumping is environmentally okay, then other environmental uses for sludge have to be evaluated on a feasibility and cost basis. Unless there is some net beneficial effect, yet to be determined, from putting New York City sludge on Pennsylvania strip-mines, we shouldn't do it."

Wagner believes that emerging government regulations and local public opinion will make land disposal increasingly difficult and unattractive. "I think we should be pushing for more ocean dumping," he says, "not less."

Hardly had Wagner spoken than other sludge producers were heading for the high seas. The first attempt to crash the exclusive 106-Mile Club came from the genteel City of Boston. The Athens of the New World is daily dumping 600,000 gallons—70 tons—of awful offal, salted with a significant *soupçon* of toxic industrial wastes, into beautiful Boston Harbor, near where the quintessence of clam chowder comes from. On notice from a federal judge to clean up its act, the Massachusetts Water Resources Authority has applied for permission from EPA to share the deep-water dump off New Jersey. Christopher J. Daggett, New York regional administrator of the federal EPA, stands ready to entertain Boston's application "as the last resort," if he can be persuaded that Boston has no alternative land site available and that ocean dumping is the "best option."

So far no one is worrying about what EPA will do in five years, when its designation of the 106-mile site expires and all questions are reopened. Until then, at least, the sludge will continue to sink out of sight and out of mind.

TROUBLED WATERS[3]

The last day of the summer season dawned gray and drizzly on the New Jersey coast. But instead of rushing for home, many departing vacationers went down to the sea. In some beach communities they formed small knots, in others lines more than a mile long as they silently linked hands and stared out at brown-tinged waves. They had come to pray for the ocean.

For those thousands of tourists, resort owners, and residents on the 127 mi. of coast known simply as The Shore, there was little doubt that the ocean needed help. During the summer, hundreds of dead dolphins, raw sewage, tar balls, and even used syringes washed ashore. Many beaches closed because of the pollution. Local papers trumpeted cases of illnesses caught from swimming—even, some said, from simply breathing the salt spray. "The pollution has got to stop," declares Linda S. Hasbrouck, who heads "Save Our Shores," the local environmental group that organized the Labor Day event. "It's so bad you don't know what's coming in with the next tide."

Hammered Home

All along the U.S. coast, unpleasant surprises rode in on the summer surf. Parents on Massachusetts' South Shore were appalled when children began decorating sand castles with plastic tampon applicators that drifted ashore from garbage dumped in Boston Harbor. On Long Island Sound, millions of dead fish washed up—killed by a total lack of oxygen in the water. Another giant dead zone formed at the mouth of the Mississippi. Near Corpus Christi, Tex., litter from ships and off-shore drilling rigs covered the beaches at Padre Island National Seashore. Algae blooms—red and brown tides—tainted waters along the Atlantic coast.

The Adriatic and Aegean, the Mediterranean, the Baltic, the Irish Sea, and the Sea of Japan are not faring much better. In Au-

[3]Reprint of an article by Tim Smart and Emily T. Smith, *Business Week* staff-writers. Reprinted by permission from *Business Week*, pp. 88–91, 94, 98, 102, 104. O. 12, '87. Copyright © 1986 by *Business Week*.

gust an 18-m sailboat dubbed the Goletta Verde—the Green Schooner—completed a much-publicized two-month trip around Italy to assess coastal pollution. What it found wasn't pretty. From Trieste to Genoa, the waters of the Mediterranean are murky with sewage and industrial waste. Some of the world's most spectacular beaches were bespoiled by garbage. Even the legendary beauty of the isle of Capri was marred by flotsam from nearby Naples. The Bay of Naples was fouled with such "incredibly high levels of raw sewage that it's on the outer limits of the imaginable," says Claudio Pirro, a marine biologist from Rome's Weltman Laboratories, who made the voyage.

This summer of 1987 hammered home a point that some scientists have been making for years: The world's coastal waters are in trouble, deep trouble. The seas' capacity to absorb a lethal cocktail of industrial, urban, and agricultural wastes is being exceeded. And when those overstressed ecosystems are exposed to natural insults, such as unusually warm weather, they collapse. "All our coastal systems are damaged, some so badly that we can't use them anymore," sighs Joseph A. Mihursky, a professor at the University of Maryland's Chesapeake Biological Laboratory.

Ironically, concern over ocean pollution is coming on the heels of a decade of steady progress toward cleaning up U.S. inland waters. But cleaning the fouled seas is a gargantuan task by comparison, and it is likely to become one of the most pressing environmental issues of the next decade—and beyond.

While discharges from factories and municipal sewage plants are part of the problem, much of what ends up in the sea simply washes off the land—from farms, lawns, and city streets. To protect fragile marine environments, legislators will almost certainly have to tighten controls on sewage treatment and industrial discharges. But that won't be enough to cope with runoff—so-called nonpoint pollution. Everything from pet waste to crankcase oil drained in the driveway or fertilizers used on lawns contributes. Curbing it will require fundamental changes in residents' habits—and sweeping new laws that will restrict coastal development.

The price of inaction is enormous. Jolting changes in the coastal ecology affect a critical part of the food chain—and ultimately human health. Toxic chemicals and heavy metals such as cadmium and lead are picked off bay bottoms by small organisms that fish and shellfish eat. Disease organisms from raw sewage thrive. Cases of human illness, including hepatitis, from eating seafood have risen drastically since the early 1980s, according to the National Marine Fisheries Service. "Poisoning the sea will inevitably poison us," warned oceanographer Jacques-Yves Cousteau a decade ago.

Dead Oysters

The poisons in the waters already are jeopardizing the $6 billion U.S. commercial and recreational fishing businesses. In Boston's south shore suburb of Quincy, a sign on a spit of land called Hough's Neck proclaims it "The Flounder Fishing Capital of the World." No more. Today, most flounder that survive the polluted waters of Boston Harbor are hardly table grade. "If you did fish here, the first thing you'd notice is fin rot," explains Steve Hunt, director of Boston's Save the Bay—Save the Harbor. And if you gutted a fish, you'd probably find white liver tumors.

Less anachronistic are "No Fishing" signs that are cropping up all along the nation's coasts. This year a shocking 33% of U.S. shellfish beds are closed. The annual harvest from Long Island's rich clam and scallop beds, worth $110 million in the 1970s, has plunged to less than half that. The watermen of Chesapeake Bay are raking up dead oysters, killed by a mysterious disease called MSX. At Matagorda Bay in Texas, seafood wholesaler Emery Waite says: "You can't find enough oysters to make a stew."

Finfish are no different. The harvest of striped bass from New York's waters, which stood at 14.7 million lb. in 1973, is now zero. Bass fishing was banned in 1986 because the fish, which spawn in the Hudson River, were contaminated with dangerous levels of polychlorinated biphenyls (PCBs)—a toxic chemical that was dumped into the river years ago by General Electric Co. Today health officials in New York warn recreational fishermen not to eat bass, carp, sunfish, catfish, or walleye from the Hudson.

Even so, commercial fishermen continue to make money. Demand for seafood has never been higher—ironically, because many consumers consider fish healthier than red meat—and prices are spiraling upward. Last year the nation's fishing industry landed 6 billion lb. of fish, worth $2.8 billion. Meanwhile, imports have more than tripled in the past decade, to $7.6 billion last year.

The numbers tell only part of the story. In the U.S., catches are climbing because fishermen are overfishing declining supplies of some prime fish, such as flounder, halibut, striped bass, and cod. And they are keeping—and selling—the "junk" fish, such as mako shark, that they once threw back as inedible. "Supply is down, and demand is so high that the price keeps going up," says Robert Morris, who owns a 40-ft. trawler that he sails from Newport, R.I.

Yet Morris and others know that all is not well at sea. The "draggerman" used to fish in water from 30 to 50 ft. deep. But lately, Morris' nets are clogged with seaweed. He blames the nutrients from detergents for the rapid seaweed growth. So, he is venturing deeper—down to 300 ft. "Pollution is affecting how I make a living," Morris says.

Fishermen—and many scientists—believe that rampant development of the shoreline is the source of much of the pollution that threatens coastal waters. By 1990, 75% of the U.S. population will live within 50 mi. of the shore, compared with 40% in 1984, according to the U.S. Census Bureau. "We've been hellbent on development at all costs," says James I. Jones, director of the Mississippi-Alabama Sea Grant Consortium, a national program of colleges with marine studies.

The problem is that sea life likes to live near the shore as well. The deep oceans are still relatively clean, but most fish breed in the increasingly endangered coastal shallows, especially tidal marshes and estuaries where the nutrients washed from the land create a rich variety of life—and food for larger organisms. In addition, the coastal wetlands act as a natural filter to trap pesticides from farms, waste from urban runoff, and chemicals from industry.

But those wetlands are being destroyed at an alarming rate. In Florida, marinas and condominiums are replacing mangrove swamps. Louisiana is losing its wetlands at the rate of 50 sq. mi. a year to development and oil and gas exploration. At this rate, none will remain in 100 years. As the wetlands disappear, raw sewage flows unfiltered into the gulf. So since 1982 about 70% of the state's oyster beds have been closed for up to six months a year. "If you ask where you can eat an oyster and not get sick, I can't think of any place," says James Cosgrove, an official at the Louisiana Health Dept.

Toxic chemicals from coastal industry also take their toll. Even though existing environmental laws have forced industry to reduce its discharges, more than 1,300 major industrial facilities still have federal permits to dump their wastes into estuaries. Along the Texas coast, the oil industry has 2,000 permits to discharge wastewater from drilling, which contains lead, cadmium, and other toxics, into the water. Near Clear Lake, Tex., Brian W. Cain, an environmental specialist with the U.S. Fish & Wildlife Service, plunges the handle of an old fishing net into one canal's dark green water. A large bubble burps out, and a knot of oily chemicals spreads on the surface. The air quickly turns fetid. "There's nothing growing along here," Cain says.

Industrial pollution, however, is far easier to control than runoff. What happens when it rains? Wastes are flushed by the ton from city streets into the water. "Toxic runoff from the streets is a major contributor," says Gordon C. Colvin, director of marine resources at New York's Environmental Conservation Dept.

Inform Inc., an environmental research group based in New York City, compared government data on Hudson River pollution in 1982 from all identifiable sources. What it found was disturbing indeed. The amounts of 26 toxic chemicals added to the river from runoff, including PCBs and mercury, far exceeded those from other sources. Runoff contributed more than 182,000 lb. of lead, for example, compared with only 240 lb. from other sources.

The most deadly contribution from runoff is not toxic chemicals, however. It is nutrients. These can concoct the deadly "algae blooms," such as red tide, that rob the waters of life-giving oxygen. That happens because the nutrients from fertilizer or sewage cause certain microorganisms to multiply at an explosive rate, using up oxygen in the water. When they die and sink to the bottom, they decompose—sopping up even more oxygen. When levels of dissolved oxygen are very low, scientists call the condition hypoxia. When there is none at all, it's called anoxia.

'Cementified'

That's what caused this summer's fish kills in Long Island Sound. On July 21 and 22 oxygen below the top layer of water completely disappeared. Then, for the next week, dissolved oxygen levels were less than 1 mg per liter of water. Anoxia can occur naturally, of course. But Barbara L. Welsh, an associate professor of oceanography at the University of Connecticut who is studying historical data on oxygen levels, believes that "back in the days of the Indians" it was unlikely oxygen levels ever got below 3 mg per liter.

In the Mediterranean, sewage is the culprit. An estimated 90% of the sewage generated around its coast is still dumped raw into its waters. Along the French Riviera, the word for many is: Go to the beach, but don't go swimming. Yvonne P. Tilleard, a part-time caterer who has vacationed in Cannes for the past 20 years, says her children "wouldn't go near the water." She says that local pediatricians warn parents not to allow young children to swim.

In addition, the sea's European coasts are fast becoming what environmentalists call "cementified" by runaway development. "In the past few years, what virgin coastline was left has been disappearing at an alarming rate," says Costanza Pera, an official in Italy's newly created Environmental Ministry. "The risk that the whole Mediterranean basin will be ringed by cement is very real."

Nor are the seas spared the threat of radioactive waste. Low-level radioactive wastes were dumped in the 1960s and 1970s in deep waters off the coast of Spain by Britain, Belgium, the Netherlands, and Switzerland. Although a moratorium took effect in 1983, the British have recently discussed burying low-level radioactive wastes in caverns beneath the Irish Sea.

The Irish, in fact, are skirmishing with the British over a nuclear reprocessing plant at Sellafield on England's West Coast. That facility, which turns spent reactor fuel into uranium and weapons-grade plutonium and is authorized to dump a limited amount of nuclear waste into the Irish Sea, has been responsible for numerous pollution incidents since it opened in the 1950s. Just four years ago divers for Greenpeace, an international environmental group, had to be decontaminated after they discovered unauthorized radioactive discharges. Greenpeace claims Sellafield has dumped more radioactive wastes into the Irish Sea than have been disposed of in all the world's oceans combined.

Pollution is beginning to affect the open seas as well. Ships, including those of the U. S. military, always pitch anything they don't want over the rail. Overall, about 6 million metric tons of litter are tossed overboard annually.

Plastics, however, do not sink into Davy Jones's locker: They float—and they don't disintegrate. Sooner or later, anything that floats on the sea comes ashore. According to the National Academy of Sciences, fishermen lose some 136,000 tons of plastic nets, lines, and buoys a year. Tons of waste from offshore oil rigs, everything from chemical drums to food wrappers, washes up on the Gulf Coast. An international law banning dumping plastics at sea is nearing approval by the U. S. Senate.

A recent study of Laysan albatross chicks on Midway Island in the Pacific found that 90% of them had plastic residues in their digestive systems. It is estimated that 1 million birds and 100,000 marine mammals die yearly because they get stuck in plastic refuse or they swallow it and choke.

Oil pollution is a particular problem, especially along the tanker lanes from the Mideast to Europe, Japan, and the U.S. Scientists generally agree that 3 million to 6 million tons of oil are dumped into the oceans from land and sea sources each year. Most of this is not accidental. It is the residue of petrochemical operations, at-sea dumping, and general sloppiness.

Heavy Metals

Healing the seas may be an enormous task. "Once the condition of the waters deteriorates, it's very difficult to halt the decline," warns Gary Mayer, associate director of environmental studies at the National Sea Grant Program. And the path to a cleanup is enormously costly and riddled with political pitfalls.

Take Boston. Some 6,000 factories discharge oil, grease, heavy metals, and acids into its 50-sq.-mi. harbor, rendering it one of the most polluted in the country. But the city and state governments began taking steps only when ordered to by a federal court in 1983—and the bureaucracy made no move to comply until 1985. Now scientists estimate that it will take at least 12 years before Bostonians will be able to stick a toe in the water. Fishing in the harbor, now outlawed, may never come back. The cost? More than $3 billion to upgrade waste-disposal facilities, which will quadruple the water and sewage bills of the 43 municipalities that use the harbor for waste disposal.

Nationwide, it would cost an estimated $76 billion just to overhaul municipal waste-treatment plants, which produce 70% of all effluent discharged deliberately into coastal waters. Today one-third of the nation's sewage-treatment plants, including San Diego's Point Loma and Boston's Deer Island, still remove only half of all solids from human wastes. And many large cities are resisting the expense of adding so-called secondary treatment plants, like New York's North River facility, which remove 90% of the solids. In February, it took a congressional override of a Presidential veto to devote $18 billion to such projects.

Then there is runoff. Scientists are just beginning to understand how big a problem it is. And they still have no firm conclusions about what should be done to control it—or what that might cost. They suspect, for example, that it has something to do with MSX disease, which is killing oysters in the Chesapeake, but they don't know what. And so far the government has been parsimonious about handing out the money needed to answer those and other questions about complex marine ecosystems. Richard F. Delaney, director of the Massachusetts Coastal Zone Management Program, told Congress in May: "We cannot even answer basic scientific questions."

Last year the Environmental Protection Agency mounted a "near-coastal waters" program to coordinate state and federal efforts at tackling coastal pollution. But so far that effort is long on paper and short on funds. This year, in renewing the Clean Water Act, Congress allocated $12 million for a two-year study of six estuaries, including Long Island Sound and Galveston Bay. "We spend a tiny amount of money trying to protect the natural environment when you place it against what is spent detracting from it," says Craig R. O'Conner, regional director for the National Marine Fisheries Service.

Irate Beachgoers

The summer's pollution problems have drawn the attention of key lawmakers, however. Senator George Mitchell (D–Me.) is preparing legislation that would finance a study of pollution in the Gulf of Maine. The four senators from New Jersey and New York want a similar study of the near-coastal waters of their states. Bills to ban dumping of plastics at sea and to finance marine research, meanwhile, are being considered in both houses. And Senator Frank R. Lautenberg (D–N.J.) is introducing a bill to require federal permits for garbage barges.

In addition, some states, faced with irate public beachgoers, increasingly militant fishermen, and billions in tourist dollars hanging in the balance, are taking steps on their own. This year, Maryland, Virginia, Pennsylvania, and the District of Columbia agreed to reduce the amount of nitrogen and phosphorus flowing into Chesapeake Bay by 40% by upgrading sewer plants and managing development and agricultural runoff. Some states, including Florida, Maryland, and North Carolina, have passed laws controlling coastal development.

Still, the piecemeal nature of these efforts worries some legislators, who feel the problem requires a more comprehensive approach along the lines of the 1970s attacks on water and air pollution. Representative Gerry E. Studds (D–Mass.), chairman of the House merchant marine, environment, and fisheries subcommittee, plans hearings on how to devise such a program. "To solve this problem requires an all-out, multifaceted program of prevention and enforcement," agrees Senator Lautenberg.

That day can't come soon enough for Fort Lauderdale diving-shop owner Brian Brooks. He recalls when visibility in nearby waters was 100 ft. and snapper and grouper were easily speared in off-shore waters. Now 60 ft. is considered good, and only the colorful tropical fish can be found. The snapper and grouper have fallen victim to overfishing and coastal development. "It's still beautiful, and we could still protect it," Brooks says. "But we're running out of time."

THE CASE FOR OCEAN WASTE DISPOSAL[4]

Love Canal, Times Beach, Valley of the Drums—these names are giving nightmares to every community confronted with the problem of waste disposal. Residents living near toxic waste sites are becoming increasingly alarmed about the potential health hazards, and the demand for "Superfund cleanups" is growing. In

[4]Reprint of an article by William Lahey, research fellow in the Marine Policy and Ocean Management Program at the Woods Hole Oceanographic Institution, and Michael Connor, research fellow in Health at the Harvard School of Public Health. Reprinted by permission from *Technology Review*, 86:60–68. Ag./S. '83. Copyright © 1983 by *Technology Review*.

many states, local opposition to newly proposed disposal sites has sprouted overnight, and older sites continue to fill up and close down.

With America facing a crisis in the disposal of its wastes on land, more and more eyes are turning to the sea. Many businesses and municipalities see the ocean as a cheaper alternative to building landfills, advanced treatment plants, and incinerators. They also recognize that the sea has virtually no political constituency fighting to protect its health and environmental rights. As a result, a growing number of waste generators are seeking ways to dispose of their wastes at sea. Millions of tons of waste are already being dumped into the ocean each year, and federal guidelines have failed to effectively regulate such practices.

At the same time, recent studies of marine life around specific dumping sites show that the ocean is not as fragile as we once believed. There are many unanswered questions about the impact of waste disposal on our oceans—how much is safe and at what sites. But there's little doubt that the ocean, particularly its deep-water areas, has some capacity to assimilate both sewage and industrial wastes. Today, scientific forces as well as powerful economic and political ones are pushing us into a policy that permits the use of the sea as a major dumping site.

That doesn't mean we should subject our oceans to the 10 million tons of dry sewage sludge, 30 million tons of sludge from air purification processes, 300 million cubic yards of dredge spoil, 65 million tons of fly ash, and 100,000 cubic meters of low-level radioactive waste that America's disposers, according to recent EPA estimates, would like to dump there. If ocean waste disposal is not to become a disaster of similar (or worse) proportions to the problem of disposal on land, we must first evaluate the ocean's capacity to assimilate waste and then regulate its disposal accordingly. Major changes are needed in the way we now approach ocean waste disposal. Specifically, we must devise a policy that relies on economic incentives to limit environmental and health risks. One approach would be to charge a fee for ocean waste disposal according to the amount of wastes to be dumped and their level of toxicity.

Protecting the Clear Blue Sea

But before we can discuss the future, it is important to put the past into perspective. In the late sixties and early seventies, the American public seemed determined to protect the ocean. The generally heightened awareness of environmental issues during this period was only one of the catalysts for concern over ocean dumping. Public ire had also been aroused by the dumping of a large amount of nerve gas into the ocean by the U.S. Army in 1970. In the same year, the President's Council on Environmental Quality (CEQ) released a report concluding that stringent legal measures, both nationally and internationally, were required to protect the ocean. The CEQ reasoned that uncertainty regarding the environmental effects of ocean dumping should inspire caution, and its report concluded that until proven harmless, the dumping of materials that appeared environmentally sensitive should be discontinued.

Congress took its cue from the public. On a single day in 1972, it passed into law three major marine-protection bills: the Marine Protection Research and Sanctuaries Act, the Marine Mammal Protection Act, and the Coastal Zone Management Act.

In 1972, Congress also sought to control piped discharge of wastes into local rivers and estuaries. It amended the Federal Water Pollution Control Act, making it unlawful to discharge a pollutant into water without a permit from the Environmental Protection Agency (EPA). The amendments set an ultimate goal of eliminating all discharge of pollutants into navigable waters by 1985. They also required many communities to build secondary sewage treatment plants by 1977. In contrast with primary treatment, which uses only mechanical means such as screens to remove suspended solids from wastewater, secondary treatment uses microorganisms to break down organic compounds in waste. The objective is to decompose these compounds sufficiently so that when the liquid effluent is released into the nation's waters, no further oxygen is required to complete the process of decomposition; depleted oxygen levels can depress fish growth and survival rates. To encourage communities to build expensive secondary treatment facilities, the federal government held out an alluring carrot: it would pay 75 percent of the construction costs and the rest was to come from state and local funds.

The nation's protective attitude toward the ocean was also mirrored in the EPA's efforts to regulate the dumping of industrial waste and sludge—the solid material left after sewage has been treated. For instance, EPA officials imposed a 1981 deadline for the termination of all ocean dumping of municipal sludge and industrial waste. The agency was successful in stopping the dumping activities of more than 100 small municipalities, but those dumpers accounted for only about 3 percent of the total municipal waste dumped in 1978. The EPA was more successful in its efforts to control the dumping of industrial waste. Between 1973 and 1980, the volume of industrial waste dumped in the ocean was cut by almost half—from 5 million tons to a little over 2.5 million tons. The EPA also took a strong stand against the dumping of low-level radioactive waste: no permits were issued for this type of disposal.

More Expensive by Land

Today, however, the public mood and the nation's regulatory climate are very different—as is the economic picture. Over the last 10 years, the costs of land-based disposal systems have soared for two primary reasons: the myriad environmental regulations and the diminishing availability of suitable sites. While many of these regulations are necessary to protect human health and the environment, they have made it difficult for some municipalities to convert to new disposal methods. New York State, for instance, has imposed a two-year moratorium on using sludge as an agricultural fertilizer out of concern that crops might absorb cadmium, a heavy metal, from the soil. A sewage district in Salem, Mass., recently spent $3 million to build an experimental incinerator facility to dispose of its sludge. But a test burn unexpectedly revealed that the resulting ash would be classified as hazardous waste under EPA regulations, and there are no hazardous-waste-disposal sites in Massachusetts. The city subsequently filed for and received a permit to discharge its waste into the ocean. The incinerator stands idle.

High energy costs have also made incineration a relatively expensive proposition. The recently required use of air-pollution-control technology has further increased the expense. Primarily because of high operating costs, nearly one-fifth of all sludge incinerators constructed since 1970 have ceased operation.

Furthermore, rising land costs as well as opposition from local residents are forcing many businesses and municipalities to transport their wastes to isolated sites farther and farther away. The resulting rise in transportation costs has motivated many disposers to look for cheaper alternatives.

As prices for disposal of waste on land rise, the cost differential between land and ocean disposal grows. Representatives from Orange County, which encompasses Los Angeles, recently told a congressional committee that the cost of discharging the county's sewage (both effluent and sludge) into the ocean was one-fourth the cost of disposing it on land. Ocean disposal, they testified, costs between $13 and $21 per ton, whereas landfilling would run about $82 to $92 per ton. Boston, which seeks to continue discharging its sewage into the ocean under a special EPA waiver, has estimated that ocean disposal costs between two and nine times less than other alternatives. New York City has estimated that the land-based alternatives for disposing of its sludge are about 10 times as expensive as its current practice of hauling the sludge 12 miles out to sea and dumping it.

The Path of Least Resistance

As mentioned, public opposition to land-based waste disposal also presents major obstacles. In the mid-1970s, for instance, Nassau County, Long Island, which has dumped its sludge into the ocean since 1963, was told by the EPA to stop dumping. The county then constructed a $14 million composting facility, only to be dissuaded from using it because of intense local opposition to spreading the resulting compost on land. Ironically, there has been no evidence suggesting that land spreading of composted sludge would contaminate groundwater. But Nassau County continues to dump its waste into the ocean, and county officials are considering using the multi-million-dollar composting facility as a parking garage.

Some EPA regulations on ocean disposal have also been successfully challenged in court, preventing the agency from maintaining its strict policy of protecting the ocean. In 1980, for instance, a number of municipalities in California and Alaska sued the EPA, asserting that it was being overly restrictive in issuing waivers to the requirement that sewage be given secondary treatment. These municipalities argued that building and operating costly secondary treatment plants was unnecessary, since they could discharge their wastes into active ocean waters that rapidly assimilate and disperse it. The Washington D.C. Circuit Court agreed, in part, and directed the EPA to allow these and other municipalities to apply for permits to dispose by sea. Since the court decision in 1982, more than 200 municipalities have applied for such permits. The EPA is now in the process of reviewing the flood of applications.

The city of New York also mounted a successful court challenge of EPA regulations in 1981. City officials believed that the EPA had acted unreasonably because its 1981 deadline for terminating all ocean dumping had not taken into consideration the costs as well as the health and environmental risks of land-based alternatives. A federal district court agreed, ruling that EPA could not impose such a deadline without considering such factors.

In a highly unusual move, the EPA chose not to appeal the decision. Former EPA administrator Anne (Gorsuch) Burford said she wanted to let the ruling stand because it gave her agency needed flexibility. But U.S. Representative Norman D'Amours (D-N.H.) called her decision "a betrayal of congressional trust" and a "devastating blow to efforts to end harmful dumping practices." Having chosen not to appeal, the EPA is faced with the task of revising its now-obsolete regulations for ocean waste disposal. How they will be rewritten under the new administration of William Ruckelshaus, who was named to succeed Burford in May, remains to be seen.

Cuts in federal spending have also dampened the effort to protect the nation's waters. Under the Reagan administration, for instance, federal spending on the construction of sewage treatment facilities has declined by almost half—from $4.2 billion in 1979 to $2.4 billion in 1982.

Faced with mounting public opposition, increasing costs, and a haphazard regulatory environment, America's businesses and cities are taking the path of least resistance. Cities such as Philadelphia, which had been forced to end its ocean dumping, have indicated an interest in returning to the sea. And other cities that have never dumped before are now exploring the possibility of applying for an ocean dumping permit. In the meantime, the number of tons of municipal and industrial waste already being dumped in the ocean is on the rise. More than 7 million tons of sludge, for instance, were dumped in 1982, compared with approximately 4.8 million tons in 1973. Researchers at the National Oceanographic and Atmospheric Administration (NOAA) have estimated that the amount of sewage sludge dumped could increase by nearly 150 percent (to 17 million tons) by 1987.

The amount of material dredged from the bottom of harbors and channels and dumped in the ocean is also increasing. Between 1977 and 1979, for instance, the amount of dredged material dumped nearly doubled—from 41 million cubic yards to approximately 73 million cubic yards. Moreover, as deep channels are constructed to increase access to U.S. ports over the next few years, the amount of dredged material to dispose of will increase. The planned expansion of existing coal ports alone may require dredging and disposing of more than six times the amount dumped in 1979.

There has also been a surge of interest in dumping low-level radioactive wastes into the ocean. In January of 1982, the U.S. Navy announced it was considering dumping decommissioned nuclear submarines in the ocean. The Navy says more than 100 aging nuclear submarines will have to be disposed of at a rate of 3 to 4 annually over 30 years. Meanwhile, the Department of Energy is interested in the ocean disposal of soil contaminated by radionuclides as a result of the Manhattan Project during World War II. Under Burford, the EPA said it was planning to revise its regulations to permit ocean dumping of low-level radioactive wastes. Afraid the EPA might be moving ahead too quickly, Congress enacted a two-year moratorium on such dumping. That deadline is up next spring.

The Effects on Marine Life

A number of recent scientific studies have been used to buttress the argument that the ocean should be used as a dumpsite. But on closer inspection, these studies confirm the need to approach the disposal of wastes at sea with a cautious and soundly developed management policy.

The most comprehensive series of studies on the impact of ocean dumping on marine life has focused on the coastal areas around Los Angeles and San Diego. For more than a decade, five municipalities and counties in Southern California have discharged their sewage effluent and sludge into coastal waters. In the last two to three years alone, an average of a billion gallons of sewage effluent a day has been discharged there. The studies were conducted by the Southern California Coastal Water Research Project (SCCWRP) and funded primarily by the five local governments who use the dumpsite. The scientists surveyed algae, fish, and plankton as well as species of worms, clams, and crustaceans that live on the ocean bottom in these areas, comparing them to marine life in areas with no known pollution.

The researchers found both positive and negative changes. While some organisms grew bigger and more abundantly near the areas of discharge, other species disappeared or suffered from a greater incidence of disease. For instance, the scientists found significant fin erosion among the flatfish that lie on the bottom near the piped discharge. They also noticed thinning in the shells of eggs laid by pelicans living on a nearby island and a resulting decline in their population; the scientists attributed these abnormalities to the discharge of DDT. More recently, the amount of DDT and other chlorinated hydrocarbons being discharged in municipal waste has declined owing to more stringent EPA regulations; the incidence of fin erosion and abnormalities in pelican eggs have declined as well.

In studying a major dump site on the East Coast, investigators also found evidence of significant degradation of marine life. The area they surveyed is located in the New York Bight apex in relatively shallow water (about 20 to 50 meters deep) 12 miles southeast of New York City. The investigators, who were affiliated with NOAA, the University of Rhode Island (URI), and the State University of New York at Stony Brook, found the animal communities living on the sea bottom were severely altered. As in Southern California, the flatfish suffered from fin erosion, and most fish species were contaminated with polychlorinated biphenyls (PCBs) and other highly toxic compounds. In response to these findings, the state of New Jersey recently issued a fish advisory suggesting that people not eat locally harvested fish more than once a week.

As early as 1970, the Food and Drug Administration had banned shellfishing within 11 kilometers of the same dumpsite after finding evidence that shellfish were contaminated with coliform bacteria. These bacteria come from human excrement and often indicate the presence of agents of infectious disease. In the late seventies, other investigators found further evidence of coliform contamination. However, the New York Bight apex is polluted from a number of sources, so it is not clear what portion of these effects can be ascribed to the dumping of sludge.

Scientists at the Woods Hole Oceanographic Institution, URI, and NOAA also studied the impact of dumping industrial waste at a deep-water site much farther offshore. Here they found little evidence of degradation of marine life. Since the late sixties, manufacturers have used the site, located 106 miles east of New York City in waters about 2,000 meters deep, to dump wastes that contained sulphuric acid, ferrous sulphate, and small quantities of other metals. Because of the EPA crackdown and the recent recession, the amount of industrial wastes dumped dropped from about 2 million tons in 1973 to 800,000 tons in 1982.

Studying the site in the late seventies, the researchers found that increased concentrations of pollutants were temporary and confined to the immediate area of dumping. These concentrations were smaller than those found to be toxic in the laboratory, and the investigators were unable to conclusively attribute the changes they found in marine life to the material being dumped. The researchers think that the seeming absence of short-term effects is due to the dynamic currents of the ocean in deep-water areas. As Woods Hole biologist Judy Capuzzo notes, ocean waters at this site have a much greater capacity to dilute wastes because of the proximity to Gulf Steam currents and the depth of the water. Wastes are often diluted to almost a millionth of their original concentrations within a few days.

The problem is that most of today's ocean disposal is taking place in areas much less equipped to handle wastes—such as the Boston Harbor and New York Bight apex, which are shallow and less active sites. Most municipal and industrial dumpers are reluctant to discharge waste at deeper sites further offshore for one simple reason: cost. It's much cheaper to dump wastes a few miles offshore than to haul it hundreds of miles out to sea.

Obstacles to Change

Overall, these results indicate that the ocean has some capacity to assimilate waste—particularly active, deep-water areas fairly far from shore. But the question of how much waste it can assimilate and at what specific sites has not yet been answered. And these answers may be difficult to come by, not only because of political and economic obstacles but also because of some scientific stumbling blocks. The fact is that our thermometer for measuring the effects of waste disposal is not very finely calibrated. We have not yet developed the tools with which to predict the ocean's capacity for waste. We haven't decided, for instance, which organisms to study: "pollution-sensitive" species such as amphipods, crustaceans that are among the first to die in a polluted area; "ecologically important" species such as zooplankton that are essential to the marine food chain; or commercially harvested species. We also haven't determined which effects should be used to assess environmental impact: changes in a particular organism's enzyme system; changes in its ability to grow and reproduce, the death of an individual organism; or the death of its entire population. We also haven't decided which area should be used to calculate damage—the dumpsite boundary itself, the zone of initial dilution from a specific dump, or some larger, as yet undefined boundary.

Furthermore, many of the methods now used to test toxicity in the lab tell us little about toxicity in the ocean. In the lab, researchers usually test different concentrations of chemicals on a particular marine species, calculating safe levels through repeated experiments. But we can't conclusively link the results of these laboratory tests to actual effects on organisms in the ocean.

It is also difficult to separate the effects of waste disposal from other environmental perturbations caused by humans. For instance, the region around the New York Bight apex receives pollutants from river flows, atmospheric fallout, landfill leaching, and the dumping of dredge spoils, dirt, and sludge. It is hard for scientists to determined whether a decrease in a particular fish population is due to sludge dumping or to air emissions from automobiles and industry.

Some scientists believe that a threshold for safe levels of waste can be determined for ocean dumping much like the threshold that is used for waste disposal in freshwater. According to that standard, fish can survive only in waters where the dissolved oxygen content is above 6 parts per million; disposers are prohibited from discharging wastes that will lower the oxygen content below that level. While this particular threshold has been effective in areas where simple organic wastes are discharged, much of today's sewage contains a variety of industrial contaminants. Consequently, a number of different thresholds will probably be required to accurately predict the level of danger to marine life.

Furthermore, developing thresholds for ocean dumping depends on how we define unacceptable consequences. For example, a coastal area will be able to assimilate more waste if we define our threshold as undesirable effects on commercially harvested species rather than undesirable effects on the marine ecology. And if we decide to concern ourselves only with edible fish, then we must determine what kind of change we consider unacceptable: the reduction, say, in a particular fish population of 1 percent, 10 percent, or 50 percent.

The Risks by Land versus Sea

In recent years, many scientists, politicians, and industry representatives have suggested we are being overly protective of our oceans at the expense of our groundwaters. In two separate reports, one released by the National Advisory Committee on Oceans and Atmosphere (NACOA), a presidential committee, and the other by the National Academy of Sciences (NAS), authorities have voiced the concern that our stringent policy of protecting the ocean is only increasing the danger to public health by forcing the disposal of all wastes on land. Some NACOA and NAS scientists expressed particular concern over the contamination of groundwater from the leakage of PCBs and other hazardous compounds dumped in lagoons and landfills.

There's no question that toxic wastes are seeping into groundwater and soil from dumpsites throughout the nation. But how do these potential hazards compare with the risks of dumping wastes into shallow water offshore?

Unfortunately, very few studies have attempted to answer that question. Perhaps the most extensive information comes from Nassau County, Long Island, an area where both fish and groundwater risks are high. Offshore lies the New York Bight, the nation's largest dumpsite for solid sludge; onshore, sandy soil and a particularly shallow aquifer combine to make groundwater contamination a problem as well. Based on data available from Nassau County officials, we have been able to compare the health risks posed by the daily consumption of locally harvested fish versus the daily consumption of groundwater.

First of all, the types of organic compounds that contaminate groundwater and fish are quite different. Small, soluble chemical compounds such as trichloroethylene and tetrachloroethylene are usually found in groundwater, having leaked from dumpsites; both are common industrial solvents. Conversely, large compounds such as PCBs dissolve poorly in groundwater and tend to cling to surface soils, but they are easily adsorbed and stored by fish. As a result, their concentration in fish is tens of thousands times their concentration in water. Both types of compounds are suspected carcinogens.

We have calculated the total risks of getting cancer from both types of contaminants by multiplying the degree of their potency by their estimated dose in humans. For an average individual weighing 155 pounds, the additional lifetime risk of getting cancer from drinking 2 liters of water a day is 34 in a million—a level of risk three times greater than the 10-in-a-million threshold considered acceptable by the EPA. For the same individual, the increased risk of getting cancer from eating 6.5 grams of fish per day (a mixture of flounder, lobster, and mussels averaged from a normal weekly portion) is 65 in a million. So even though most people drink much more water than they eat fish, the carcinogenic risks of doing either are about the same. For people who consume a large amount of striped bass, the risk of getting cancer is much higher: 2100 in a million. (Striped bass are more toxic because they migrate into the highly contaminated waters of the Hudson River.)

According to this limited assessment, we should be just as concerned about dumping chemical pollutants such as PCBs in shallow ocean waters as we are about dumping them on land. In fact, it might be safer to treat many of the insoluble organic compounds in active land-based systems. On the other hand, the ocean, with its large dilution capacity, is better equipped to handle wastes such as metals. And of course, some industrial pollutants such as dioxin and PCBs shouldn't be dumped at all. It would be much safer to destroy these compounds by incineration or develop substitute products or processes that would eliminate the need to dispose of these highly toxic substances.

An Economic Incentive for Safety

All these studies point to one conclusion: we need to make major changes in the way we regulate ocean disposal. The protectionist approach of the 1970s has backfired. Exceptions to restrictions on ocean dumping were granted and deadlines extended haphazardly, often as a reaction to political and economic forces. The United States must develop a regulatory program that recognizes that the ocean should be used for some waste disposal, yet provides an incentive for businesses and towns to control their ocean dumping activities.

One approach we advocate would be to charge a sliding fee for ocean disposal based on the amount dumped, the type of contaminants in the waste, and the location of disposal. Since it is far cheaper to dump wastes in the ocean than on land (no one, after all, can buy or lease the ocean for dumping), a fee system could bring ocean disposal into economic parity with other alternatives. Decisions on where to dispose of waste would then be more likely based on comparisons of environmental and health risks, not on the basis of economic or political expediency.

Under a fee system, disposers of innocuous materials such as seafood-cannery wastes, which contain easily degradable compounds, could be charged low, if any, fees, since uncontaminated organic material poses few threats to the marine environment. Disposers of contaminated wastes, on the other hand, would be taxed according to the type and concentration of pollutants in the waste. This kind of graduated fee would create an incentive for dumpers to either reduce the volume of waste dumped or pretreat it to decrease the amount of contaminants present. For instance, a municipality that was being charged a relatively high dumping fee because of the high concentrations of industrial contaminants in its waste would have an incentive to impose pretreatment requirements on local industries. Indeed, the Clear Water Act already gives municipalities the legal authority to do so.

A variable fee system would also give industry an incentive for developing innovative pretreatment and disposal techniques. The less contaminated the waste a company discharges, the lower the fee would be. Both industry and government could also save on fees (and make some additional income) by processing sludge into fertilizer and other useful products.

A variable fee system could also be used to encourage dumpers to use appropriate dumpsites. Under this system, dumping at deep-water sites shown to have a greater capacity to absorb waste would be charged much less than dumping at shallow coastal sites that are already heavily polluted. The fees could be structured in such a way as to counterbalance the increased cost of hauling wastes further offshore to deeper, more appropriate sites.

Furthermore, revenues generated from dumping fees could be put to valuable use. They could be earmarked for research of alternative waste-disposal techniques and used to finance studies aimed at finding safe deep-water dumping sites. Such revenues could also be used to fund a comprehensive program to monitor dumping and assess its effects on the marine environment. Since our knowledge of the marine environment is rudimentary, any effective system of ocean waste disposal must be accompanied by an ongoing monitoring program.

Finally, we believe a variable fee system would be more equitable than the current regulatory system. For example, the EPA phased out many small-volume municipal dumpers during the 1970s while allowing large-volume dumpers (such as New York City) to continue without penalty. The large municipalities had the legal and financial resources to find loopholes in the system, forcing the EPA to grant them special exceptions. A fee system based on tonnage and contamination levels would eliminate these inequities. In short, we believe such a system is a versatile regulatory tool that deserves serious consideration by America's policymakers.

OCEAN DUMPING[5]

Simplicity is the most deceitful mistress that ever betrayed man.
                                                    —HENRY BROOKS ADAMS
                                                         (1838-1918)

This statement summarizes the past situation in regard to
ocean dumping. The oceans have immense volume. They are dynamic and have powerful dispersal forces. A simple extrapolation
of these facts has appealed historically in those segments of society searching for a means of waste disposal. The dilution and dispersal forces at work in the oceans, coupled with a capacity to
degrade or deteriorate many materials, seem to point to the
ocean for waste disposal with few accompanying adverse effects.
The attractiveness of this idea is illustrated by recent compila-
tions of major ocean dumping locations and total amounts
dumped over the period 1976-1979.

The contentious issues of ocean dumping revolve around the
question: What is the capacity of the ocean for receiving wastes
without adverse effects? Logically, the answer to that question depends on the composition of the waste, the proposed disposal site,
the duration of the disposal, and the definition of an adverse effect. Within that context, we seek to review briefly the history of
ocean dumping in the United States and make some recommendations for the future.

Why Was Ocean Dumping Restricted?

In the decade of the 1960s and into the early 1970s, research
clearly demonstrated that man-made toxic wastes could be detected in the farthest and deepest reaches of the oceans. The
quantities detected were very small: 10^{-6} to 10^{-12} grams per gram
of marine organism tissue or oceanic sediment. However, the
chemicals detected, such as DDT, PCB, and the radioactive fall-

[5]Reprint of an article by John W. Farrington, director of the Coastal Research
Center at the Woods Hole Oceanographic Institution; Judith M. Capuzzo, Associate
Scientist at the Institution's Biology Department; Thomas M. Leschine, Policy Associate at the Institution; and Michael A. Champ, Professor at the American University in Washington, D.C. Reprinted by permission from *Oceanus*, 25:39-50. Winter
'82/'83. Copyright © 1983 by *Oceanus*.

out from nuclear weapons tests, had been in use only 10 to 40 years. Given the vastness of the oceans, the rapid invasion by even trace quantities of these chemicals was viewed with concern by several scientists, because laboratory experiments and field observations in near-shore areas had demonstrated real or potential toxic effects across a wide variety of marine biota.

Concurrently, there were a few incidents of toxic chemicals entering the coastal areas of the oceans and becoming a health hazard to man. For example, in Minamata Bay, Japan, mercury in a chemical plant's effluent entered the bay in sufficient quantities to pollute shellfish and fish to the extent that many people who ate these organisms became seriously ill, or died. (See *Oceanus*, Vol. 24, No. 1, p. 34.)

A second example, not as tragic, involved contamination of fish by polychlorinated biphenyls (PCBs). These organic chemicals, whose production and distribution was banned in the 1970s, were used in electronic components, such as capacitors and transformers, mainly since the 1950s. They leaked into the environment in a variety of ways and eventually entered the oceans. Fish caught in some coastal areas contained high enough concentrations of PCBs, in a few cases, to cause a significant number of reproductive failures when fed to minks over a period of time. Although obviously there are wide differences between minks and people, the fact that both species are mammals understandably caused concern about the release of PCBs to the environment and subsequent adverse effects on humans. Other more extensive evidence eventually led to a ban on production of PCBs in several countries and restrictions on PCB use in others.

There were several additional incidents that led to increased concern about ocean pollution. Of course, there also were numerous examples of disposal of wastes in the ocean for which no threat to man nor overt adverse impacts were noted; however, the few documented adverse impacts served as warnings of problems in the future. The fact that some bodies of water smaller than oceans had serious problems (the Thames River in Britain, the Houston Ship Channel, and Lake Erie, to name a few) bolstered the arguments for caution against waste disposal in the ocean. For some, the cries of "polluted lakes today, the North Atlantic tomorrow" or "the oceans are dying" were simple extrapolations. The combination of a few concerned scientists and environmentalists, a larger body of concerned citizens, and concerned elected and appointed officials was sufficient to begin the process of regulating and limiting waste discharge in the oceans.

New laws were passed by the U.S. Congress, and rules and regulations were promulgated and implemented. These laws built upon earlier marine pollution control laws dating back to the Rivers and Harbors Act of 1899.

Retreat from a Ban on Ocean Dumping?

In the 1976 to 1978 period, a few engineers and scientists began to realize that, while marine scientists were pointing to the real and potential effects of ocean disposal of urban sewage sludge, among other materials, other engineers and scientists dealing with the pollution of rivers, lakes, and groundwater were successfully advocating construction of advanced sewage treatment facilities. However, no one had planned adequately for the disposal of the new sludge.

In 1977, Congress adopted a 1981 statutory phase-out deadline for sewage sludge that might cause unreasonable degradation of the marine environment. This placed New York City, in particular, on the horns of a dilemma. Where could New Yorkers dispose of the sludge they had been dumping in the New York Bight? As the 1981 deadline approached and economically reasonable alternatives for New York's sludge disposal were not at hand, there was considerable socio-economic and political pressure to re-examine the ban on ocean dumping. Several court cases evolved in regard to New York City's sewage sludge disposal in the New York Bight and the criteria by which sludge could be judged as acceptable or not acceptable for disposal at a given site in the ocean.

During this same period, the United States citizenry was awakened to the seemingly more immediate problem of festering hazardous waste disposal sites on land. We cannot recount here the many examples, such as Love Canal. It suffices to state that the problems with land disposal made it seem ludicrous to ban unequivocably ocean dumping.

In July of 1979, a group of scientists, engineers, and a few observers met at Crystal Mountain, Washington, to re-examine the issue of waste disposal in the oceans. After lively debate and agonizing writing and rewriting, a report, "Proceedings of a Workshop on Assimilative Capacity of U.S. Coastal Waters for

Pollutants," was issued. Although there were many important cautionary statements and caveats in the report, the principal message was that in certain circumstances the oceans probably could be used as receivers of waste without undue harm to the oceans or to man. At the very least, said the group, the issue should be examined in context with the other alternatives for waste disposal. The National Advisory Committee on Oceans and Atmospheres (NACOA) reached a similar conclusion two years later after again examining the issues.

Testimony during May, June, September, and November of 1981, and in March of 1982 before the House Subcommittee on Oceanography, the Subcommittee on Fisheries and Wildlife Conservation and the Environment, and the Committee on Merchant Marine and Fisheries addressed the contentious issue of waste dumping. John A. Knauss, Acting Chairman of NACOA, summarized the problem in his testimony, which identified a legislative crisis in regulating the disposal of wastes:

Five federal statutes affect the management of society's waste material. They are: Federal Water Pollution Control Act (FWPCA, often referred to as the Clean Water Act); Marine Protection Research and Sanctuaries Act (MPRSA, often referred to as the Ocean Dumping Act); Safe Drinking Water Act (SDWA); Resource Conservation and Recovery Act (RCRA); and the Clean Air Act. It was impossible to implement all five statutes simultaneously and as a result the implementation of each shifted the burden of receiving society's waste products to the medium that was least regulated at the moment. An industry or municipality faced with the problems of what to do with its waste may well find that the Clean Air Act effectively prohibits incineration, the FWPCA and the Ocean Dumping Act similarly limit disposal at sea, and the RCRA and the SDWA effectively prohibit land disposal or deep-well injection. Based on our review of this history and the statutes, NACOA is concerned that this medium-by-medium approach has produced groups of regulations whose primary objective is to protect a particular medium from its use as a waste disposal medium without any regard for the impact of these regulations on other media.

The Present Situation

Most of the concern about ocean dumping and ocean sewer outfalls is derived from the presence of toxic chemicals, viruses, and pathogenic bacteria in much of the waste discharged to the ocean. Viruses and pathogenic bacteria enter sludge in sewage treatment plants as a result of processing human and animal wastes. Known or potentially toxic, mutagenic or carcinogenic chemicals, such as PCBs and chlorinated pesticides (DDT and chlordane, for example) can enter sludge as a result of rain washing material from the atmosphere or by dust falling from the atmosphere to city streets. The chemicals are then transported to and through sewers by rainwater runoff. Factories add still more toxic chemicals directly into municipal sewer systems. In older East Coast cities, such as New York, the problem is compounded by combined storm and sanitary sewer systems which convey normal rainfall to sewage treatment plants and also flush some untreated sewage directly into nearby ocean waters when there is an overflow.

Sewage sludge is mostly the remains of solid wastes and a mass of bacteria that has been degrading some organic products from human and industrial wastes. In simple terms, the sludge sewage treatment process is a larger, more sophisticated version of a backyard septic system. As with septic systems, bacteria in the plant periodically need refurbishing. The old bacterial mass is removed. Along with this mass of bacteria, there are recalcitrant chemicals, such as PCBs, many chlorinated pesticides, and some petrochemicals, that are difficult to break apart or degrade biologically. Many of these chemicals are not very soluble in water and are adsorbed onto the surfaces of bacteria during the treatment processes. Other chemicals interact with the sludge in chemical reactions, which results in their removal from the sewage as it passes through the treatment plant. The net result is sludge containing elevated concentrations of toxic chemicals.

Environmental concern with sludge disposal is focused on 1) the accumulation and transfer of these toxic chemicals in marine food chains, 2) the toxic effects of such chemicals on survival and reproduction of marine organisms, and 3) the uptake and accumulation of pathogenic bacteria and viruses in commercially harvested species destined for human consumption.

Dredging spoils from urban harbors, rivers, and estuarine areas often contain elevated concentrations of pollutant chemicals because of industrial and municipal sewer discharges to these areas and runoff from land carrying the fallout from urban air pollution. The pollutant chemicals are present in elevated concentrations in both sludge and dredged harbor sediments as the result of the initial attempt to release the chemicals to the environment for dilution to innocuous concentrations.

Chemical wastes from industrial chemical operations also are dumped at sea and are candidates for increased ocean dumping in the future. Again, the problems revolve around the toxic portion of the waste, which is often a very small part of the total mass of material.

Aside from the greater concern about toxic chemicals and pathogens, there are still concerns about arbitrarily releasing degradable organic matter and nutrients to the ocean. If these substances are discharged in high enough amounts to some oceanic areas of poor dispersion and mixing energy, then depletion of oxygen in the area as a result of so much microbial degradation of organic matter may become a threat to some species of commercial importance. Eutrophication of coastal areas from nutrient enrichment may result in changes, both in the types of species that live in a given area and in the dynamics of marine food chains, with consequent loss of commercial resources. For example, during the summer of 1976, poor water circulation and high nutrient inputs to the inner continental shelf off northern New Jersey resulted in a bloom of the dinoflagellate *Ceratium tripos* off the New Jersey coast. This species of algae is rarely used as a food source by zooplankton; thus, a great deal of organic matter not utilized by marine food webs was left to decompose. This resulted in a high rate of oxygen consumption by bacteria decomposing this organic matter (the dead algae), creating hypoxic, or low-oxygen, conditions in near-bottom waters and on the ocean floor. These adverse conditions caused mass mortality of fish and shellfish and commercial losses.

On the other hand, it may be possible, under certain controlled-release conditions, to stimulate the biological productivity of an area and increase the yield of valuable seafood. For the majority of the cases, we suspect that the disposal process itself will be of sufficient social or economic benefit so as not to warrant the extra effort required to demonstrably increase a yield of seafood. The emphasis is now (and will be for several years) on preventing adverse effects, which carry with them social and economic costs.

Waste Management Strategies

The give and take of the Congressional hearings on ocean dumping during 1981 and early 1982 illustrates that we are in a critical period of transition from a regulatory stance that was approaching a ban on ocean dumping to . . . what? The scientific, engineering, and political debates are intense (see *Oceanus*, Vol. 24, No. 1). Policy-makers who govern regulatory actions may be about to embark on a more rational course, toward multi-media (air, land, sea) assessment prior to decisions about where to dump or discharge wastes in the future, though many important issues remain and may require a decade to be resolved.

There are five general waste management strategies, with several options within each category. The costs and benefits of each option should be evaluated and should enter into the decision on how each type of waste is managed. This appears to be a rational and straightforward approach. It *is* rational. However, it is *not* straightforward, because much of the knowledge required to make meaningful cost-benefit comparisons is not available.

REDUCING AND RECYCLING WASTE

The Global 2000 Report prepared for President Jimmy Carter provided a sobering look to the future. A growing shortage of materials of various types will occur as we proceed toward the year 2000. Thus, we expect that incentives to conserve will increase. There is a growing movement already in the United States toward recycling in local neighborhoods and communities, encouraged nationwide by some manufacturers. The recent documentation of reduced energy utilization, and thereby reduced fossil fuel consumption, may be a further indication of such a trend. We think these are indications that society can adjust its life-style in a relatively short time.

During the 1970s, the production of several toxic chemicals was reduced. The restricted use of DDT and a few other chlorinated pesticides, and the ban on PCB production in the United States, followed a few decades or less after scientific evidence emerged suggesting these chemicals caused environmental damage. Perhaps the alternatives were less effective and more costly, but the important point is that adjustments have been made. Thus, if the evidence is compelling and if the environmental damage is extensive enough or potentially extensive or, more importantly, if human health is at risk, then action can be taken and generation of toxic wastes can be reduced.

Ideally, the removal of many toxic chemicals from sewage sludge could be achieved by keeping the material out of sewers in the first place. Controls on industrial effluent releases to municipal systems have been in force or proposed for several years. The 1977 amendments to the Clean Water Act require that communities seeking waivers from wastewater secondary treatment requirements develop programs by which their most toxic industrial wastes are removed from municipal wastewater. There is debate about the application and expense of new technologies to reduce industrial chemical releases. Some people argue that the costs to consumers or loss of jobs in a given region make the application of effluent controls untenable. Certain toxic chemical inputs to sewers will be decreased in the 1980s, but there will still be enough industrial effluent input to significantly contaminate many urban sludges.

Another problem of equal or greater significance is the fact that many chemicals of concern enter sewers via dispersive release to the environment or because they are already used extensively by society. Two examples illustrate this. Some polynuclear aromatic hydrocarbons are known mutagens and carcinogens. These compounds enter sewers as a result of chronic dribbling of oil from industrial operations and automobile crankcases. They also are released to the atmosphere during combustion of fossil fuels and deposited on the ground by dry fallout or by rain, and then washed into sewers.

The second example concerns PCBs. Even though they are no longer produced in the United States, a significant amount of PCBs are still in use. Burning of PCB-containing electrical components in municipal incinerators or leakage from electrical components in use releases PCBs to the environment. A portion of this release is collected in sewers via atmosphere deposition and runoff. Thus we cannot look to controls on effluent releases from industrial plants to completely solve the problem of toxic chemicals in sludge.

Furthermore, a significant problem with waste disposal in the oceans is related to contaminated sediments from dredging operations near urban areas. This is a problem of relocating toxic chemicals already released to the environment. Dredged material is a major ocean dumping input to the ocean. The 7 to 10 million cubic yards of material dredged annually from New York Harbor is sufficient to cover the borough of Manhattan six inches deep. At present, the proportion of dredged material worldwide that contains concentrations of chemicals of concern is not known, but most dredge spoils from industrialized harbors are heavily contaminated.

INCINERATION OF WASTE

There is a growing conviction among scientists, engineers, and officials of regulatory agencies that high-temperature, high-efficiency combustion offers the best means of disposing of certain very hazardous chemical wastes. The Environmental Protection Agency (EPA) has allowed several test burnings of chlorinated organic chemical wastes at sea where the basic chemical nature of seawater rapidly neutralizes the hydrochloric acid that is the main combustion product of concern. Burning such wastes on land requires difficult and potentially expensive controls on the release of this acid to prevent adverse effects on nearby structures, plants, animals, and people.

It appears that the incineration method could be extended to sediments polluted with high concentrations of toxic organic chemicals, using specially designed rotary kilns on ships. However, this technology is only in the early prototype stage.

The main concerns about adverse effects associated with this treatment strategy are 1) ensuring continued high efficiency of operation; 2) preventing accidental spills of material during collection, storage, loading, and transit at sea; and 3) the cost of the fuel necessary to achieve the required temperatures. Current forecasts indicate that this strategy will be economical and of best use to society when applied to low volumes of highly toxic materials.

DISPOSAL ON LAND

Most of the sewage sludge currently generated for disposal in the United States is disposed of on land. Less than 15 percent is released in the oceans by ocean dumping or by ocean outfalls. Nevertheless, the difficulty of allocating sufficient land for landfill disposal operations, spray irrigation, or land spreading of composted sludge material is a major obstacle to the land disposal option, especially near urban areas where land is more expensive and more wastes are generated. The second problem with land disposal is the protection of public health. Contaminated groundwater and polluted air plague some land disposal sites, and thus are of potential concern for all sites.

Spreading of composted sewage sludge on agricultural lands or spraying treated sewage in forests has been researched and is in use in several inland locations, particularly in the Midwest.

However, public health concerns related to the presence of pollutant chemicals and pathogens have prevented more widespread use. Also, suitable agricultural lands and forests are not always within easy reach of urban areas, where the bulk of waste is generated, especially in coastal areas. Thus the costs of transportation and land discourage the use of this option.

In the United States, many cities are near the ocean. A recent EPA study estimated that 25 percent of all sludge generated for disposal comes from counties that border the ocean. The costs of using an area of ocean for disposal are not appreciable at present, compared to the costs of land near urban areas. Thus, ocean disposal has continued appeal; it releases land for other uses.

DISPOSAL IN THE OCEAN

There are two basic man-made modes for delivery of wastes to the oceans: pipes and ships (or barges). The engineering aspects of each are not germane to this discussion. However, there are some fundamental decisions that need to be made as to the mode used and the location of the release.

The two most prevalent scenarios (simplified here) of ocean disposal are near-shore disposal and deep ocean (far away from land) disposal.

Near-shore disposal. The arguments in favor of this disposal scenario are:

Recoverability. If a mistake is made in estimating the severity of adverse effects, then it should be technologically easier to recover the wastes for alternate treatment or disposal than if the material had been disposed in the deep ocean. This argument presumes that the disposal area is a low-energy environment—an area where mixing and turbulence by waves, currents, and storms will not significantly disperse the material.

Impact could be restricted in area. Again, if a low-energy environment is used, then it might be possible to sacrifice a small area of extreme adverse impact in order to minimize effects elsewhere. This is the near-shore equivalent of the landfill disposal option.

Impact could be minimized by dispersal in a high-energy mixing area. The argument here is that tidal and wind-driven currents, storms, and wave-induced turbulence, prevalent in some coastal areas, provide the extensive mixing needed for initial dispersion and dilution.

Research monitoring and management of waste. It is easier and less expensive to conduct research and monitoring to verify predicted adverse effects and discover whether unsuspected, unwanted adverse impacts are about to occur, or have occurred. This presumes our knowledge about coastal and estuarine processes is more advanced than for the open ocean and that it is easier to monitor and conduct research in coastal areas.

Economic. Near-shore disposal is cheaper, often by a factor of three or more, than deep ocean disposal. More importantly, at a time when costs for all types of disposal are increasing rapidly, near-shore disposal is frequently much less expensive than any other disposal alternative for coastal communities. According to one wastewater treatment estimate, sludge handling, transportation, and disposal now account for 35 percent of capital costs and 55 percent of annual operation and maintenance costs.

The disadvantages of near-shore disposal are:

Proximity to people. Disposal sites are close to land and population centers. If an adverse impact is discovered, there will be less time to protect human health.

Proximity to valuable living resources in coastal areas. The reasoning of the preceding point applies here as well.

Neither high-energy mixing areas nor quiescent areas are always near the activities generating the wastes.

Deep-ocean disposal. The proponents of deep-ocean disposal generally cite the following:

Extensive dilution and dispersion. Concentrations of toxic materials can be diluted in a large volume, thereby minimizing effects on marine ecosystems. Such extensive dilution also makes it less likely that the material will return to man in harmful amounts. Similarly, the exposure to contaminants is reduced for the living resources of near-shore coastal areas.

The disadvantages are:

Recoverability may be impossible. If it is determined that unwanted adverse effects are in progress, then there will be a serious problem. Despite the resolve and great technological ingenuity of society for solving difficult problems, recovering dispersed wastes from the open ocean will be a nearly impossible task and certainly disruptive to the economic well-being of the nation.

Research and monitoring are difficult. The volume of oceanic areas involved and the limitations of our present knowledge make it difficult to check on what is actually happening.

Economics. The transportation of wastes to the open ocean is more expensive than for the coastal option.

International complications. While inputs to coastal areas can eventually reach open ocean areas, the direct dumping of pollutant chemicals in open ocean waters beyond the limits of the contiguous zone could affect the waters of neighboring countries and contaminate living resources fished by nonadjacent countries. For example, ocean dumping at the 106-mile site off New York could seriously contaminate squid, which are fished near that area by Japanese fishermen and marketed for human consumption in Japan. Ocean dumping also is regulated by the London Dumping Convention. The U.S. is a contracting party to that convention and should adhere to the ban on ocean dumping of certain toxic materials.

Focus for the 1980s

We need to contend with the following essential problems if we are to proceed on a rational course with respect to ocean dumping:

Time and Space Scales of Extrapolation. We need to better understand how to extrapolate from experiments and field observations of short duration (weeks to months) and small spatial scale (meters to kilometers) to the fate and effects of waste disposed for decades in areas the size of the entire northeastern United States coastal area or the Southern California Bight and nearby continental slope.

Unreasonable Degradation. We need further knowledge of the long-term effects of individual chemicals, bacteria, and viruses in various wastes in the marine environment. What constitutes unreasonable degradation of the marine environment and how do we predict such damage? From a scientific viewpoint, assessing the effects of pollutants on the marine environment is a difficult task. It requires an understanding of the adaptive and disruptive responses of each level of biological organization, from cellular to population responses, and how those responses may affect responses at the next level of organization. Only when the compensatory or adaptive responses of the cell or whole organism begin to fail, do deleterious effects on the population become apparent. For predictive purposes, one must be aware of the early warning signs of stress before compensatory mechanisms are surpassed.

From a socio-political viewpoint, the level of environmental degradation acceptable to the public may be quite different than that accepted by some of the scientific community. For example, filling in of the Hackensack Meadowlands in New Jersey to build a sports complex resulted in irreversible damage to a salt marsh in the view of an ecologist but the creation of a valuable recreational facility in the view of a New York Giants fan. Future decisions on the effects of ocean dumping will require an integration of scientific, social, and political viewpoints.

Comparisons of Land and Sea Options. What is the common scientific language by which we can compare effects on a forest ecosystem subjected to decades of waste disposal with effects on a fishery subjected to the same waste input for the same period of time? How can these comparisons be communicated in a meaningful way to those engaged in policy and regulation, and to the public? The degree of damage, the recovery time necessary to restore a natural community, and the resiliency to further damage must be considered in any comparison of ecological systems subjected to waste disposal.

How can the proper comparative studies of specific disposal options for specific materials be conducted when government responsibility is fragmented among several agencies and spread over several levels of government? Most studies to date have focused on single disposal options or the effects of disposal on a single disposal medium.

Flexible Policy. A major misconception about ocean dumping is the expectation among the regulators and the public that a decision is forthcoming, based on solid scientific evidence. They also think that, once the decision is made, we can get on with investigating society's other problems. Such thinking minimizes the complexity of the issues and exaggerates the ability of science to provide predictions that will stand the test of time. We recognize that there has been important and exciting progress on local, regional, and global scales in understanding environmental processes. However, there is still much research that needs to be undertaken and completed.

Of equal importance is the fact that decisions on the future of ocean dumping cannot be made on the basis of scientific and technical information alone. Regulators must be prepared to make important decisions on questions of appropriate societal values: how much pollution of land or water is to be tolerated for the sake of economic, health, or other social benefits that accrue because alternative disposal options are not taken? Given the uncertainties about scientific and technical facts, and the changing nature of societal values, can a policy be implemented that is flexible enough to incorporate changes as a result of value shifts or new scientific information? If a marginally tolerable level of ocean pollution is to be accepted for the sake of economic or other considerations, does the policy selected include sufficient incentives for ocean disposal users to generate better waste management methods in the future?

We are concerned that, once the decision to continue or increase ocean dumping is made, there will not be the follow-up in continued research and monitoring which is required to assess the accuracy of estimates made about the fate and effects of the wastes. We caution that such estimates and predictions are often no better than predictions of the behavior of the economy of the United States. Economists are allowed to continue to collect data to revise and update their predictions and assessments because society recognizes that there are many uncertainties in economic predictions. Scientific assessments of many aspects of waste disposal in the oceans are of a similar nature, and society should accept that, too.

Research and Monitoring

Much of the required information for making decisions can only come from fundamental research. The remaining information comes from monitoring what happens at a given site. Ideally, there should be a five-to-ten-year period of study at a few dumpsites and disposal areas with the various characteristics previously described. In fact, studies of some of the options have been under way for several years in the New York Bight, the Southern California area, and Deep Water Dumpsite 106 off the northeastern U.S. coast. A few more studies like these, incorporating revisions based on lessons gleaned from earlier work, are essential.

There is very great danger of the "snowball made into an avalanche" effect with respect to ocean dumping, which should be avoided at all costs until reasonable assessment of the options is completed. The problem is that if one municipality or industry is allowed to use the oceans for waste disposal, then many others may cite the precedent and follow. One recent estimate is that municipal sludge dumping could increase by as much as 150 percent if all municipalities that could exercise the dumping option actually did. The situation then could get out of hand before the required data and assessment supportive of extensive dumping is available.

Understanding the fate and effects of materials discharged to the oceans depends on our fundamental knowledge of oceanic processes. At a time when society is poised for a massive experiment with the oceans—continued and possibly substantially increased ocean dumping—fundamental ocean research, and even research and monitoring applied to marine pollution studies, is being severely curtailed. Oceangoing research vessels are being decommissioned and tied up at docks. Even many near-shore research projects require an understanding of open-ocean processes. For example, waves that provide mixing energy in coastal areas are often generated in the open ocean. Many research and monitoring projects on the continental shelf and in areas such as the New York Bight and the Southern California Bight require the larger research vessels to safely handle gear and carry enough scientists to study efficiently and synoptically several facets of the problems. Furthermore, we cannot rely too much on studies of coastal organisms when considering effects on open-ocean organisms. Sensitivities to pollutants are known to vary by as much as factors of 10 or more when comparing organisms for these two different oceanic regimes.

Certainly, remote sensing from aircraft and satellites is a tool of growing significance in ocean research. However, this is no substitute for most ship-based work. Our warning is explicit. If ocean dumping is to be a viable option for waste disposal, then United States ocean scientists have to be able to get to sea to make certain that estimates (many times "guesstimates") of adverse impacts (or no adverse impacts) are correct. There also has to be a concomitant increase in stable research funding in order to understand such important topics as what governs natural fluctuations in marine ecosystems, and to recognize the early indications of changes induced by man's activities.

United States activities in ocean waste disposal will be increasingly watched by other nations and could become part of foreign policy interactions. Wastes discharged to the oceans are not contained by political boundaries. The failure of the United States to support the Law of the Sea Treaty could have repercussions if the U.S. ocean disposal policy in its Exclusive Economic Zone differs from standards agreed upon by other nations. Some nations may well protest United States activities, while others may adopt policies leading to less rigorous standards for disposal of wastes in the ocean. We should be cognizant of the ocean disposal plans of our nearest neighbors in Canada and Mexico, and evaluate the total input to our contiguous oceanic areas. Likewise, if the United States engages in disposal of wastes in open ocean areas, an evaluation of potential long-term impact *must* take into account the activities of other countries that may release wastes into the same or contiguous areas.

For these reasons, we anticipate the 1980s will be a period of vigorous national and international debate about ocean dumping. The debate could extend well into the 1990s.

Summary

We agree with those who argue that the oceans have a capacity to receive some wastes without undue harm to valuable living resources or to public health. We also agree with those who are concerned that a mad dash toward using the oceans as a less expensive, quick fix for waste disposal will occur without due consideration of the relative risks and benefits of all options for waste management.

There are encouraging indications that those who formulate policy and promulgate regulations are evaluating all options. However, we are discouraged that this policy evaluation will probably acknowledge the essential role of continued research and monitoring activities in the rational evaluation and implementation of the various options, while not providing the means for those activities to be effectively carried out. We do not advocate unnecessary delays in decisions about where to put waste today. Rather, we advocate a flexible policy which explicitly recognizes that today's decision can be re-evaluated, modified, or even abandoned as new knowledged is acquired.

One of the most serious mistakes that could be made about disposal of wastes in the ocean is to decide now that we have sufficient knowledge to establish regulations for 20 years or even 10 years. We do not. Our present knowledge of the oceans teaches us how complicated oceanic processes can be; it is rudimentary compared to the questions asked. In the lexicon of computer buffs, "garbage in, garbage out" is not a good way to decide where to put the garbage.

III. HAZARDOUS WASTE

EDITOR'S INTRODUCTION

Waste disposal also includes the disposal of hazardous substances, which affect public health and safety and have become the source of considerable controversy and debate. Perhaps no aspect of waste disposal has made so many newspaper headlines. The Love Canal case in 1978 brought about the creation of "Superfund" by Congress in 1980, to be administered by the Environmental Protection Agency (EPA). A $1.6 billion, five-year crash program, it was designed to clean up thousands of leaking dump sites threatening to contaminate underground water supplies. In five years, however, the EPA accomplished almost nothing. Of nearly a thousand hazardous waste sites slated for cleanup, only 6 were cleared—and these not fully or adequately. Although a cleanup is now underway, it is estimated that the task may require $23 billion, and many feel that such an estimate is low.

In the first article in this section, "A Problem That Cannot Be Buried," Ed Magnuson, a staffwriter for *Time*, reviews the background of the Superfund and the faltering attempts of the EPA to begin cleaning up toxic waste sites—efforts that have been further hampered by budget cuts by the Reagan administration. In "The Midnight Dumpers," in *USA Today* magazine, journalists Judith and Mark Miller reveal the widespread involvement of the "mob" in illegal dumping. In many cases, companies engage carting concerns to dispose of their hazardous waste for them, and these contractors then merely dump drums of toxics into the soil or disguise the discharge as refuse and leave it at city dumps, where it will leak into the soil and groundwater.

In a related article in *Forbes*, James Cook focuses upon the growth of legitimate waste disposal companies that charge large fees to meet EPA standards, and handle disposal on both land and sea. As he notes, a growth industry in toxic waste disposal now involves ocean-going incinerators, which eliminate problems of landfill. EPA estimates, indeed, that it will need a fleet of over thirty incinerator ships by 1990. In another article, in *The Nation*, journalists Andrew Porterfield and David Weir report on the surreptitious transport of domestic waste to Third World countries. Particularly disturbing is their finding that one of the principals in this shadow industry is the U.S. government itself, particularly the military. Turning to another aspect of waste disposal, James J. Holbrook, in *Sierra*, explores advances in biological technology in dealing with toxics—bacteria that "eat" dioxin and other harmful substances. Finally, Fred H. Tschirley, in an article in *Scientific American*, raises the question of how harmful dioxin actually is to human beings. Deadly in even minute amounts to experimental animals, it has not been shown, he contends, to have severe effects on humans when exposure is extremely marginal.

A PROBLEM THAT CANNOT BE BURIED[1]

The great toxic-waste mess oozed its way into the nation's consciousness, and its conscience, a little more than five years ago. "An environmental emergency," declared the Surgeon General in 1980. "A ticking time bomb primed to go off," warned the Environmental Protection Agency. The reaction was typically all-American: Congress created a grand-sounding "Superfund," a $1.6 billion, five-year crash program designed to clean up thousands of leaking dumps that were threatening to contaminate much of the nation's underground water supplies.

Last week that law expired, a victim of wrangling among the Senate, the House and President Reagan over how much more should be dedicated to the cause and who should pay the bill. During its existence, the Superfund dribbled away most of its money on a mismanaged effort that served only to reveal the almost unimaginable enormity of the task ahead. Though Congress is likely to reach an agreement by next month on a new infusion of money, anywhere from $10 billion over five years (the House proposal) to $5.3 billion (the Reagan Administration's figure), for now the once ambitious program lingers in limbo.

[1]Reprint of an article by Ed Magnuson, *Time* staffwriter. Reprinted by permission from *Time*. 126:76–84. O. 14, '85. Copyright © 1985 by *Time*.

Meanwhile, fears about toxic wastes continue to grow. Each day more and more communities discover that they are living near dumps or atop ground that has been contaminated by chemicals whose once strange names and initials—dioxin, vinyl chloride, PBB and PCB, as well as such familiar toxins as lead, mercury and arsenic—have become household synonyms for mysterious and deadly poisons. "The problem is worse than it was five years ago," contends New Jersey Democrat James Florio, who as a Congressman from one of the most seriously contaminated states became the key author of the 1980 Superfund law. "It's much, much greater than anyone thought." Concedes Lee Thomas, the third director of the scandal-tarnished EPA during the Reagan Administration: "We have a far bigger problem than we thought when Superfund was enacted. There are far more sites that are far more difficult to deal with than anybody ever anticipated." That comes as no surprise to Barry Commoner, the venerable environmental gadfly. Says he: "We are poisoning ourselves and our posterity."

The growing awareness of the vast scope of the toxic-waste problem has bred much public anguish but precious little remedial action. The Office of Technology Assessment, a research arm of Congress, contends that there may be at least 10,000 hazardous-waste sites in the U.S. that pose a serious threat to public health and that should be given priority in any national cleanup. The cost, OTA estimates, could easily reach $100 billion, or more than $1,000 per U.S. household. Eventually, predicts the General Accounting Office, which also does studies for Congress, more than 378,000 waste sites may require corrective action. So far the EPA has put only 850 dumps on its priority list. In its five-year effort, it managed to clean up only six sites and, critics protest, not very thoroughly at that.

The U.S. faces other grave environmental risks: acid rain, smoggy skies, radioactive wastes and lethal gases escaping from industrial plants. Over the past five years, the EPA reported last week, mishaps in the handling or production of chemicals have caused some 1,500 injuries and 135 deaths. But the disposal of dangerous wastes is clearly the most pressing concern. Toxic dumps where steel drums have been left to rust and leak, letting poisons seep into the earth for decades, are scattered in virtually every county of every state. They present a potentially irreversible threat to water supplies, public health and the economy.

Why has so little been accomplished in attacking the chemi-
cal-dump mess? "If we're looking for people to blame, well, the
woods are full of them," says William Ruckelshaus, who helped
launch EPA as its first director during the Nixon Administration,
and who was recalled by Reagan in 1983 to try to repair the agency's image.

Most critics direct their anger at the current Administration.
William Drayton, chairman of a Washington-based environmen-
tal-safety group, says it took "an enormous movement in American history" to develop a national consensus that "this country is
going to provide public health protection against chemical
contaminants." But what followed, charges Drayton, who served
as assistant EPA administrator under President Jimmy Carter,
was "a classic Greek tragedy: enter stage right the Reagan revolution with its enormous ideological antagonism to regulation of
any sort. You have a leader who just doesn't understand what all
those Latin-named chemicals are and what they do. On this subject he just stopped learning." Says Douglas Costle, Carter's EPA
director, of his successors at the agency: "They just flat-out didn't
realize they had a tiger by the tail until it bit them in the ass."

In the Administration's defense, Ruckelshaus argues that
"the Government had never done anything like it before, starting
from absolute scratch to deal with this terribly emotional mix of
issues. The fact that there were mismanagement, false starts and
mistakes was inevitable." But even he admits that the agency's
performance on toxic wastes "didn't have to be as bad as it was."

There is little doubt that EPA has seemed feckless and con-
fused. For one thing, its critics contend that less than 20% of the
original $1.6 billion Superfund allocation has been spent on actu-
al cleanup of waste sites. The National Campaign Against Toxic
Hazards, an umbrella group of grass-roots activists, claims that
less than 10% of the 850 sites on EPA's current priority list have
received any remedial attention at all in the program's first five
years. At that pace, according to the group, "millions of Americans will wait decades for the EPA to clean up their poisoned
communities."

Apart from actual cleanup, EPA is responsible for monitoring
sites suspected of endangering underground water supplies, so
that the citizens who draw such water can be warned of the health
dangers. But a congressional research team concluded last April
that of the 1,246 hazardous-waste dumps it surveyed, nearly half

showed signs of polluting nearby groundwater. The EPA's monitoring of these sites, the study charged, was "inaccurate, incomplete and unreliable."

The congressional watchdogs claim that when EPA finally does tackle a waste site, it seeks only a stop-gap solution to the chemical seepage. When a dump is cleaned up, its wastes are often merely shifted to other locales, "which themselves may become Superfund sites," the OTA report says. "Risks are often transferred from one community to another and to future generations."

A poll taken last month for *Time* by Yankelovich, Skelly & White, Inc., shows that 79% of Americans say that "not enough" has been done to clean up toxic-waste sites. More surprising, when asked, "Would you be willing to pay higher state and local taxes to fund cleanup programs in your area," 64% answered yes (34% said no, 2% were unsure).

This attitude toward the slow pace of dump cleanups is part of a broad public sense that Government is failing to respond adequately to environmental concerns in general. Some 45% of those polled said that current laws to protect the environment do not go far enough, while 29% are satisfied with them and 16% think they go "too far." Fully 63% feel that even the inadequate governmental protections are not being enforced strictly enough by the agencies involved.

Some critics contend that putting off the admittedly expensive cleanup effort will mean greater expense in the future. "Delay not only prolongs the time that people are exposed to toxic hazards," says Michael Podhorzer, director of the National Campaign Against Toxic Hazards. "But every day it means that more toxic chemicals are released into the soil, air and water. The longer we wait, the greater the damage will be and the higher the final cleanup cost will be."

Consider the meager six sites deemed to have been cleaned through the Superfund. After a nine-month-long spill of chemicals into the Susquehanna River starting in 1979, it was found that a small Pennsylvania company had been systematically, and illegally, dumping toxic wastes into shafts that fed into the Butler Tunnel, an outlet for waste water from abandoned coal mines near Pittston, Pa. Three men were convicted of violating the state's Clean Streams Act, and one was sent to prison. The three and their company were fined $750,000. EPA supervised the cleanup of the river pollution, and in 1982 it took the site off its priority list. But heavy rains from Hurricane Gloria sent 100,000 gal. of oily, smelly chemical wastes rushing back up to the surface of this presumably cleaned-up site and into the Susquehanna. "There was an extremely strong odor that would burn your nostrils," said City Clerk Paul McGarry, who went to investigate after residents began phoning with complaints. "It looked like liquid tar."

Another of the six sites that EPA claims to have successfully cleaned is in Baltimore, where strong acids and *aqua regia*, one of the most corrosive liquids in existence, had been stored throughout the 1970s. For years, residents in 20 row houses along Annapolis Road complained of eye, nose and throat irritation; eight people were burned in July 1979 when chemicals leaked into a playing area. EPA removed 1,500 drums and scraped off up to twelve inches of topsoil. The land was sloped and sodded and declared fit for a playground. But critics cite tests showing that the contamination had worked its way as far as 15 feet below the surface. No attempt was made to prevent seepage of these deeper chemicals into the groundwater or a nearby river.

For some 40 years, beginning in the 1930s, the Velsicol Chemical Co. (formerly the Michigan Chemical Co.) had dumped and burned toxic industrial chemicals on a 3.5-acre site along the Pine River near St. Louis, Mich. A county golf course was developed beside the dump. By the mid-'60s, fish in the river contained high levels of such known or suspected carcinogens as PBB, PCB and DDT. Working with EPA, the company in 1982 agreed to spend $38.5 million to clean up the area. At the golf course, all soil was removed to a depth of 3 ft. below any signs of contamination. That involved hauling 68,204 cu. yds. of dirt away. Fully 1.25 million gal. of contaminated groundwater were pumped into a 3,400-ft. well lined with two cement walls. EPA considers the golf course cleaned up, as indeed it seems to be. In one sense, however, the problem was merely transported across the river. All that soil has been deposited on the plant's property, where a bigger cleanup job has been completed.

Two of the six sites chosen by the EPA for quick action should probably not have been on the top-priority list in the first place. In Greenville, Miss., Walcott Chemical Co. had stored 226 drums of such chemicals as tetrasodium pyrophosphate and formic acid in a warehouse that the state of Mississippi had seized for failure to pay taxes. The state considered the chemicals a fire hazard (rather than a contamination threat) and asked EPA to put the site near the top of its list. The agency merely had the drums hauled off to an approved landfill in Emelle, Ala. Problem solved. Similarly, about 700 drums of chemicals had been stored in a Cleveland warehouse used by Chemical Minerals Recovery Co. Another 700 were piled outside the building. None had sprung significant leaks. But EPA gave the site priority and had the drums carried to an EPA-licensed landfill in Geneva, Ohio. Another site cleared.

However, what about the residents of Emelle and Geneva? Have they inherited the old headaches of Greenville and Cleveland? Perhaps not immediately, since the dump at Emelle sits atop hundreds of feet of clay, and the one at Geneva at least has the now mandatory clay liner. But most experts consider any landfill only a temporary solution to the chemical-waste problem. Eventually, all will develop cracks or gradually give way to the corrosive action of the potent chemicals.

Six California environmental groups recently surveyed seven landfills in that state. Though the EPA was monitoring them for leaks, the groups reported, "every one of the sites examined is leaking, without exception; and every one is out of compliance with currently applicable regulations." Wastes placed in them from other failed sites may soon have to be picked up and moved once again. The result is a bleak game of chemical leapfrog.

The spreading realization that there is no easy way simply to bury the toxic-waste problem has fed the ever present NIMBY (not in my backyard) syndrome. "Something's got to give," protests Christopher Daggett, EPA administrator for New York and New Jersey. "Either we aren't going to have cleanups, or someone's going to bite the bullet and start accepting wastes. But Lord knows, no one wants to be first." Daggett and his boss, EPA Director Thomas, contend that there is no ready technology that can promptly solve the disposal problem. "We can't wait around until we have the ultimate answer," says Daggett. "This stuff is still being generated, and we have to deal with it today. So, yes, we are going to put it into landfills that may leak someday. But give me an alternative. Do you want me to store these wastes in drums all over the country?"

Critics accuse EPA of being too cautious in failing to rely more heavily on such destruction technologies as high-temperature incineration and in failing to back innovative approaches for detoxifying chemical wastes. EPA has projects under way in these fields, but the pace is slow, the funding inadequate, and there is little sense of urgency. Barbara Vecchiarelli, a citizens'-group leader in Marlboro Township, N.J., admires Daggett's dedication to his work but, nonetheless, complains about EPA in general: "They don't have the technology to handle chemical pollution. The problem is bigger than they are, and they're afraid to admit it to the American people."

Part of the problem with EPA's management of the Superfund over the past five years stems from Reagan's initial choice of top officials who were ill-prepared to handle the difficult mandate. Anne Burford, a Colorado lawyer and Republican Party fund raiser, was tapped in 1981 to head EPA; at White House urging, she approved the selection of Rita Lavelle, a California publicist who had worked for a chemical company (Aerojet General Corp.), to direct the Superfund start-up. In the mismanagement that followed, Lavelle was convicted of perjury for denying any involvement in EPA's dealings with the Stringfellow Acid Pits, a notorious waste dump in California, where Aerojet General, along with many other companies, had dumped tons of caustics, cyanides and heavy metals over the years. Burford was also charged with contempt of Congress for refusing to give it some internal EPA documents; the charge was dropped after she quit in March 1983.

While EPA was floundering, the White House imposed drastic funding cuts, resulting in the loss of 23% of its budget and 19% of its employees by 1983, even though the toxic-waste work load was multiplying. When Ruckelshaus was named EPA chief after Burford's resignation, he managed to rebuild the staff's morale, restore some of its funding and give his successor a stronger hand. Thomas, the former head of the Federal Emergency Management Agency, has worked hard at getting EPA into gear. "He kicked the tires and punched the fender and said, 'Let's get this thing moving,'" notes former EPA Chief Costle.

But where is EPA going on toxic wastes? "We've learned that we have a far bigger problem than we thought when Superfund was enacted," Thomas concedes. "But we have a good bit of momentum now. I don't see how anybody could come into this agency and run it faster than we've tried to run it over the past couple of years." He can point to the fact the EPA is currently trying to stop the spread of pollution at about 200 sites and is preparing to tackle a cleanup of about 200 others. By next year, he predicts, "we'll be managing nearly a thousand sites at the same time."

The new emphasis at EPA has been logical enough: stop the seepage of pollutants and protect drinking water first, get rid of the toxic stew later. When a number of wells in Sag Harbor, on New York's Long Island, were found contaminated, EPA moved swiftly to have two dozen affected homes hooked up to a city water system. "The longer-term problem isn't solved," says EPA's Daggett. "But we were able to remove the immediate threat." In 1981 poisons were discovered in 27 of 30 wells serving Battle Creek, Mich. An elaborate system of purge wells was created to pump the contaminated water out of its underground plume and purify it. Now Battle Creek has 16 clean wells from which to drink.

Finally, acting on a law passed by Congress in 1976, EPA has issued tough regulations designed to trace the flow of toxic chemicals from their manufacture to their eventual disposition, creating a paper trail that should discourage illegal dumping and pinpoint responsibility when contamination occurs. The agency has also vastly tightened its licensing requirements for anyone operating a landfill that is permitted to accept hazardous wastes. By early next month, all such landfills can continue to operate only if they have double liners to prevent seepage. Already, wells must be bored in the surrounding area to detect any signs of spreading contamination. Although clearly a necessity, the new rules may take half of the nation's 2,000 licensed disposal sites out of operation, further aggravating the problem plaguing chemical-waste planners: nowhere to go.

A prime example of a modern disposal facility is the one operated by Waste Management, Inc., at its C.I.D. Hazardous Waste landfill in Chicago. A giant excavation 35 ft. deep covers two acres. A floor of compacted clay approximately 40 ft. thick has been laid below the bottom of the hole. On top of this virtually impermeable bed, workmen are placing a plastic liner to be topped by a plastic-grid system that will collect and direct any seepage into a series of sump pumps. Above the grid will be another plastic liner, another layer of clay and yet more plastic. A plumbing system will pump rainwater out of the area. Nearby, the company is spending $1.6 million to improve its large surface collection tanks, made of concrete lined with epoxy, that receive waste from steel-processing plants. New fiber-glass liners are being placed inside the cylinders. In the past, such wastes were merely poured into noxious surface lagoons. (In other ways, Waste Management is no ideal disposer. It agreed to pay $2.5 million last April to settle an EPA charge that it had illegally disposed of toxic chemicals in Ohio.)

Such techniques are, of course, expensive. But the increasing cost of getting rid of dangerous chemicals provides a powerful incentive for manufacturers who use them to find ways to recapture and recycle them. While Government pressure and supervision of toxic-waste sites are vital, the disposal problem will remain intractable unless industry does most of the job itself. By one estimate, 96% of all hazardous wastes never leave the property of the companies that produced them.

A number of companies have made some headway in curbing a generation of the poisons. Minnesota Mining & Manufacturing Co., for example, cut its volume of toxic wastes in half, partly by switching from solvent-based glues to water-based glues in its manufacture of adhesive tape. It also burns nearly all of the remaining wastes in a huge incinerator at Cottage Grove, Minn. "In the past five years, there has been a tremendous change in the attitude of the chemical industry about hazardous waste," says Larry O'Neill, an environmental official with Monsanto Co. in Missouri. "We are now generating less and recycling more." Still, the recovery techniques are just being developed. "When we talk about recovery, we're only talking now about 1% of all the material that's generated," claims James Patterson, director of industrial-waste-elimination research at the Illinois Institute of Technology. Even 1%, however, does add up.

If the public clamor for quicker, more effective action in the war on toxic wastes is fully justified, the expectation of easy or fast fixes is not. Some 66,000 chemicals are being used in the U.S.; EPA has classified 60,000 of them as potentially, if not definitely, hazardous to human health. They have been dumped or buried for years on the plausible but, as it turned out, tragically wrong theory that they would lose their toxicity during the decades it would take them to drift through layers of soil and rock into deep water supplies. There is no way to remedy in a few years at least a century of such misguided, if innocent, practices.

Shuffling wastes from one leaking site to another that may soon turn porous may seem absurd, but there is no way to eliminate all landfills as short-term disposal necessities. The same is true of the use of hazardous-waste incinerators. While they risk befouling the air, they are nonetheless a necessary temporary expedient.

Much more might be done, however, to find new methods of taking the poisonous punch out of hazardous chemicals. The EPA spent only $43 million in the first five years of the Superfund program on basic research and development of such techniques. According to the Office of Technology Assessment, as much as $50 million a year could be spent usefully on R. and D.

In the end, only a vast effort by the industries that profit from the chemicals can get the waste mess under control. That would undoubtedly mean added costs passed on to the consumer, but the basic fact is that the effort must be made. Wondrous chemical potions have been a great aid to mankind, easing pain, alleviating disease, prolonging life, spurring food production and serving as the catalyst for countless useful products. But once discarded, many of these concoctions, or their by-products, turn killer, and the U.S. has no choice but to curb their lethal ways.

THE MIDNIGHT DUMPERS[2]

In 1979, an Environmental Protection Agency (EPA) study identified 32,254 poisonous waste dumps around the U.S., 800 of which posed a "significant imminent hazard" to public health. Michael Brown, in *Laying Waste*, wrote: "In such a milieu, it is not surprising the U.S. finds itself today in the throes of what can only be termed a cancer epidemic."

According to Rep. John Dingell (D.-Mich.)., chairman of the House Subcommittee on Oversight and Investigations,

It is appalling enough when disposers of hazardous waste, through inadvertence or ignorance, recklessly poison the environment and endanger

[2]Reprint of an article by Judith Miller and Mark Miller, free-lance journalists. Reprinted by permission from *USA Today Magazine*, 113:60–64. Mr. '85. Copyright © 1985 by The Society for the Advancement of Education.

the public health. But it is considerably more disturbing when generators, haulers, and disposers—in order to avoid the cost of legitimate disposal—engage in the practice of illicit dumping for profit. . . . We have developed information linking organized crime to the illegal dumping of toxic substances. This comes as no surprise. In fact, it was predictable, given the lucrative nature of this activity. The involvement of organized crime in the toxic waste industry poses a continuing threat to undermine government's efforts to resolve the national problem of hazardous waste disposal.

"Midnight dumping" is a term referring to the clandestine, illegal ditching of toxic wastes. The dumpers charge law-abiding companies premium prices for proper ditching of noxious wastes, then mix the deadly cargo with ordinary trash and dump the mess at an ordinary landfill. The result is pure profit for the perpetrator and poisons oozing into the ground and water of your neighborhood. "It's so easy to mix toxic chemicals with ordinary garbage it isn't even funny," government informer Harold Kaufman, a former Teamster's official who was heavily involved in New Jersey's toxic waste industry, told the House Energy and Commerce Committee in 1980. The mob, Kaufman said, had already progressed beyond the realm of gambling and drugs and had taken over dumping of *ordinary* waste in New Jersey; hence, it was an easy step into the lucrative business of *poisonous* wastes. For two years, Kaufman, an ex-convict with gripes against several of his associates and a desire to redeem himself, wore a hidden tape recorder to meetings of mob leaders where toxic disposal was discussed. He subsequently revealed what he knew to the government and is now protected as a Federal witness living under a new identity.

Kaufman said crooked hazardous waste companies charge law-abiding companies 20 times the disposal rate for ordinary garbage, supposedly to allow for the difficult job of isolating dangerous chemicals. Instead, he testified, poisons are stored on a vacant lot somewhere, randomly dumped, or mixed with ordinary garbage and deposited in landfills. New Jersey Attorney General John Degnan said such a disposal company would demand thousands of dollars for proper disposal of a truck loaded with 8,000 gallons of toxic wastes, then take only minutes to ditch the poisonous brew alongside some dark, lonely road.

Kaufman's testimony resulted in the indictment of 57 companies—a vast conspiracy that took over poisonous waste disposal in New Jersey. Forty of the defendants pleaded guilty in September, 1982; four were convicted in April, 1983; four other defendants, including one of the biggest hazardous waste disposal companies in the U.S., Browning-Ferris of Elizabeth, N.J., and its former vice president, John M. Gentempo, received a mistrial when the jury couldn't reach a verdict. They were retried in September, 1983, and both Gentempo and Browning-Ferris were acquitted. In an Atlanta case, however, the company pled no contest to price-fixing.

Kaufman charged that Ernest Palmieri, one-time business agent of Teamsters Local 945, had been on the Browning-Ferris payroll. Palmieri has been identified by the U.S. Justice Department as a leader of the Genovese Mafia crime family.

Kaufman testified that several supposed members of the Genovese and Gambino crime families in New York ordered the setting up of a New Jersey waste-haulers association. Kaufman said he knew about the order because the presidency of the group was offered to his boss at the time, Charles Macaluso. Macaluso, who was an honorary co-host at the 1976 Democratic National Convention, was convicted on June 23, 1983, in New Jersey for illegally paying off a public official in relation to a municipal disposal contract for one of his companies, Statewide Environmental Contractors, Inc., was fined $25,000, and sentenced to a two-year prison term. He was then also a defendant in three other related cases.

In 1978, Kaufman testified, Macaluso's company held a $30,000 contract to dispose of waste sludge for the Ford Motor Co. Statewide arranged to subcontract the work to a licensed handler, the Duane Marine Chemical Co. of Perth Amboy, N.J. However, Kaufman charged, Edward Lecarreaux, owner of Duane, didn't have the necessary disposal facilities. So, stated Kaufman, Duane mixed the noxious sludge with garbage and trucked it to solid-waste landfills or rolled the barrels of contaminated filth off the company's dock into Arthur Kill. Recently, Duane Marine pleaded guilty to reduced charges in connection with the 1980 dumping of 500,000 gallons of waste into Perth Amboy sewers and paid a $25,000 fine.

State and Federal investigators found dumping operations tied to the mob in New Jersey, New York, Connecticut, Rhode Island, and Massachusetts. For example, a Bridgeport, Conn., trucking firm owned and operated by the late crime boss, Carmine Galante, dumped 6,000 barrels of poisonous explosive waste on a Connecticut farm. The same company was also involved in crooked dumping near a Rhode Island pig farm, where the poisonous brew exploded and blazed out of control for three days!

Experts on organized crime told the Senate in February, 1983, that mobsters engaged in illegal dumping of hazardous wastes were "flourishing" financially. Jeremiah McKenna, chief counsel of the New York State Select Committee on Crime, claimed there is an ongoing investigation in Manhattan into the distribution of 20,000,000 to 30,000,000 gallons of apartment heating oil that disposal companies, run by crime lords, purposely mixed with poisonous chemicals. This is a clandestine way of crookedly getting rid of dangerous chemical wastes without cost to the disposers, witnesses charged. McKenna testified that "They are . . . selling 'waste oil' as fuel in New York City containing xylene, benzene, and PCB's." Xylene and benzene are flammable; benzene and PCB's (polychlorinated biphenysl) are carcinogens. McKenna explained that, in this highly profitable, deadly game, there is payment when poisonous chemicals are taken from industrial waste generators, and additional profit when fuel oil is diluted with these waste chemicals and unknowingly burned in Manhattan's heaters.

Indeed, witnesses told the subcommittee, because the "enormous profits" in criminal waste dumping were a siren's song luring bribery and corruption, there was little law enforcement effort against "midnight dumpers" by local or state governments. At the Federal level, they charged, the EPA has been "officially blind" to the danger.

McKenna declared, "If organized crime is not rooted out of toxic waste disposal, we risk a public health catastrophe in this country." He further stated that "Observation and trained intuition lead me to forecast that organized crime's position in the toxic waste disposal industry will emerge toward the second half of this decade as an issue of profound public importance."

In *Laying Waste*, author Michael Brown notes that the entire U.S. is pockmarked with chemical dumps, with many oozing poisons into soil and water, threatening our health. Writes Brown: "We have planted thousands of toxic time bombs; it is only a question of time before they explode."

Earth Day, 1980: An Environmental Nightmare

One toxic time bomb at Elizabeth, N.J., exploded the night
before Earth Day (April 21), 1980. Explosions sent a fireball hundreds of feet skyward, and 55-gallon drums were launched into
the air, bursting into bombs. Helpless residents of Elizabeth and
Staten Island, N.Y., were shaken; 30 people were injured. The
resulting fire ravaged the area for 10 hours before it was snuffed
out. People stood transfixed in terror. Because dangerous plasti-
cizers, pesticides, and nitric and picric acids were stored at the
dump, officials closed schools in Elizabeth and Staten Island; residents were urged to stay home with their windows shut.

For years, disgusted critics had been calling the Chemical
Control Corporation's plant which exploded that Earth Day the
"Three Mile Island of Chemical Dumps." Elizabeth residents had
suffered from the stink emitted from the plant where thousands
of barrels of noxious chemicals had been illegally stashed. New
Jersey's Department of Environmental Protection removed
10,000 barrels of the explosive, poisonous materials prior to the
explosion; 24,000 barrels remained, however, and they turned
into a time bomb. A confidential Federal report found that
crooked stashing of chemical wastes and pesticides at Chemical
Control could have sent poisonous fumes, drifting like an angel
of death, over New York City!

Who ran Chemical Control? A Federal grand jury indicted
William Carracino, former president of the company, on charges
of fraud, illegal storage of poisonous chemicals, and mail fraud.
The indictment charged that Carracino defrauded four chemical
companies by not keeping his pledge to incinerate poisons legally
at Chemical Control. A separate indictment against the
company charged that John Albert, an alleged mobster, assumed
"managerial control" of it in 1977 to further his own illicit toxic
waste schemes.

Actually, assumed "managerial control" is a civilized way of
putting it. Carracino told New York State's Select Committee on
Crime that his company had literally been ripped off from him
at gunpoint by Albert. Indeed, informant Kaufman testified that,
in the summer of 1978, he received a phone call from Albert,
who, lawmen alleged, is a "soldier" in the Frank Tieri (formerly
Genovese) Mafia crime family. He stated that Albert told him
that Albert had just taken over Chemical Control and wanted to

form a chemical waste-haulers association to protect crooked territories and prices. The system, Kaufman testified, was to have been patterned after that by which solid-waste firms controlled by mobsters ran ordinary garbage carting.

John Fine, former Assistant State Attorney General in New York, also told the New York Senate Crime Committee that Albert is a "soldier" in the Tieri family; furthermore, said Fine, Albert was indicted in July, 1980, on charges of financing a crooked amphetamine lab in Plainfield, N.J.

The money behind the alleged mob takeover of Chemical Control has been traced, in part, to a wholesale fish seller in lower Manhattan. The link to the crooked fish-peddler, Joseph Lapi, a reputed organized crime figure who has since died of natural causes, emerged from investigations into the affairs of the now-defunct company. Lapi allegedly lent Albert the money to keep his Chemical Control scheme afloat. Carracino further charged that the blaze on Earth Day eve might have been sparked deliberately after the state took control of the dump from mob elements. Carracino told the New York Senate Subcommitte on Crime that a 100-gallon tank of gasoline to run cleanup equipment had been brought to Chemical Control the day of the fire, and it couldn't be found after the conflagration. Besides Albert, Carracino charged, two other men, Eugene Conlon and Michael Colleton, took over his company with guns! Carracino claimed Collecton told him: "These men—John Albert is going to run the company and you'll do as he says. Do what they tell you and everything will be all right." Carracino and Fine denounced New Jersey officials for what they said were failures to behave properly in crooked waste dumping cases.

Workers engaged in the costly cleanup after fire ravaged the Chemical Control site discovered 100 pounds of illegally stored explosives that didn't go off during the fire. The explosives, including nitroglycerin and picric acid, were kept in small containers inside 900 barrels still at the site. At the time of the explosion, 40,000 barrels had been piled six or seven drums deep in leaking stands, in violation of safety disposal codes. Officials suspect that large amounts of harmful chemicals—poisons, skin irritants, and carcinogens—oozed from Chemical Control's warehouse.

The situation is even more alarming. Were Chemical Control unique as a mob-controlled toxic waste site, it would be a blessing;

however, within the U.S., ties between toxic-waste disposers and the Mafia make a web of interlocking relationships that should be publicly unravelled. Americans have the right to know why, and by whom, they are being poisoned. What we don't know can hurt us. For instance, after the closing of the Elizabeth warehouse, a report prepared by the Trenton office of the Federal Bureau of Alcohol, Tobacco, and Firearms said: "Health officials reported that enough poisons and pesticides had been noted on the premises to provide a lethal dose to all of Staten Island and lower Manhattan in the event of a fire at Chemical Control Corporation." New Jersey officials said a year-long effort by the state prior to the fire to get rid of the dangerous wastes was the only thing which prevented this tragedy from materializing. The cleanup cost the state $10,000,000.

Interlocks between Companies

Kaufman further charged that the third largest toxic waste disposal company in the U.S., Service Corporation of America (SCA), listed on the New York stock exchange, ballooned to its present formidable size by absorbing existing carting companies, some controlled by the mob!

While an FBI probe found SCA and its former president, Thomas C. Viola, legitimate, there are alleged cases of SCA buying up companies which retained their mob links even after SCA bought them. For example, testified Kaufman, one SCA subsidiary, Waste Disposal, Inc., was purchased from Crescent Roselle in 1973, and run by Roselle until his gangland-style murder in December, 1980. According to Kaufman, such a case shows that SCA has for a long time "been involved with organized crime in the garbage business and now they're moving into hazardous waste." In other words, Kaufman claimed, crime lords, besides running illegal dump sites and fly-by-night trash-disposal companies, also kept their influence over the trash business after selling their companies to SCA.

Crescent Roselle, according to Wayne Comer, supervisor of the FBI in Newark, "attended a meeting with several high-ranking members of a traditional organized crime group"—the Tieri (Genovese) family. Roselle was an alleged associate of organized crime figures who acted on their behalf in "property rights" disputes. According to Comer,

In New Jersey, a carting company derives its value from its real property and, to a far greater extent, from its intangible property. The first type is composed of the physical property, whereas the second arises from the right to collect garbage from its customers, called stops. These "property rights" are bought and sold. . . . Because of their value, these property rights are vigorously defended. . . . Respect for property rights and the contact with or coexistence with organized crime . . . predated SCA. . . . Foremost among the companies acquired by SCA was Waste Disposal. . . . Its president, Crescent Roselle, was a powerful figure in the industry who attended grievances with other carting companies as well as with members of traditional organized crime. Before the murder of carting company owner Gabriel San Felice in May, 1978, Roselle attended a meeting with several high-ranking members of a traditional organized crime group and San Felice, at which time San Felice was ordered to return several stops to Roselle. After the murder of carting company owner Alfred DiNardi in June, 1976, Roselle met with several members of the New Jersey carting industry and an organized crime figure, at which time stops taken by DiNardi were returned to their previous owners. In general, Roselle took an active part in settling property right disputes in northeastern New Jersey as well as Ocean and Monmouth County, N.J.

Crescent Roselle, Peter Iommetti, and Ralph Mastrangelo—formerly high-level supervisors from SCA—allegedly had deep-rooted ties to the mob. Mastrangelo, owner of another SCA subsidiary, United Carting Co., was employed by SCA's staff at least until July, 1981. According to police records, he and August Vergalitto were involved in extortion from a New Jersey contractor. Vergalitto is a known associate of Tieri (Genovese) crime family member John DiGilio and John Riggi, acting head of the DeCavalcante crime family. Nevertheless, the FBI gave former SCA president Viola and his company a clean bill of health.

According to Fine, illicit toxic-waste hauling in New York and New Jersey is a mesh of interlocking concerns. He noted that, "We discovered in Orange County that a land-fill [in Warwick, N.Y., in Orange County] authorized to take only solid non-toxic waste was the site of dumping of hazardous waste materials." Trucks from Jersey were seen on the site. New Jersey forms filled in by the waste haulers "stated the wastes would be disposed in Warwick," whereas papers filed with New York "indicated disposal would be at approved toxic waste sites, not Warwick," Fine alleged. "It appeared then," he concluded, that "hundreds of thousands of gallons [of toxic wastes] were being consigned to a post office box. We had reports that toxic wastes in 55-gallon barrels, batteries, and paint matter were dumped in Warwick, N.Y."

All County Environmental Service Corp. was hauling toxic wastes to Warwick. The Warwick landfill was leased and run by Grace Disposal and Leasing Limited. The principals behind Grace were John and Frank Coppola and Louis Mongelli. Charged Fine, "Our investigation . . . disclosed that a carting company owner who underbid Mongelli's Roundlake Sanitation [an All County subsidiary] was threatened concerning the health of his family. Roundlake took over another hauling company in the area under circumstances that we wanted to investigate in the grand jury."

Toxic wastes were also illegally dumped in the Al Turi non-toxic landfill in Goshen, N.Y. Al Turi was operated by Thomas and Nick Milo and Vincent DeVito; moreover, Thomas Milo was an officer of Suburban Carting. Commented Fine, that "concern . . . was involved in . . . [a] $4,500 payoff. Subsequent arrests resulted in those cases."

Thomas Milo also owned Suburban Carting stock; Nick Milo, Thomas Milo's brother Vincent, and Vincent DeVito owned DV Waste Control Corp.; DV bought out C&D Disposal Service, a subsidiary of SCA. Before C&D was sold to DV, when Jeffrey Gaess was C&D's president, his relative, Anthony Gaess, was a defendant in the Kin-Buc toxic waste landfill case in New Jersey. "Evidence shows," alleged Fine, "numerous phone calls between Chemical Control in Elizabeth and various Gaess companies. There had been numerous shipments of toxic chemicals between Chemical Control Corp . . . and Kin-Buc."

DV took over routes of Dutchess Sanitation, Inc., Fine charged, and Dutchess was "owned and operated by Matthew 'Matty the Horse' Ianello, Joseph Fiorillo, Michael and Vincent Fiorillo. . . . A notebook entry found . . . by Federal agents . . . as an offshoot of a homicide investigation . . . is: Fiorillo friend Charles Mucillo, the telephone number 212-245-1677. That phone number belongs to one of 'Matty the Horse' Ianello's fronts in New York City. [Joseph] Fiorillo is also known to law enforcement as Joe Garbage." Ianello is also allegedly tied to criminal elements.

Charles Mucillo, an associate of Fiorillo, is a convicted gangster. His 19-year-old son was running toxic waste companies in Connecticut until they were caught illegally dumping poisons into the C. Stanton Gallops gravel pits. Charles Mucillo himself lived near the Staten Island dump site, in Travis, N.Y.

An Orange County sod farm was used to receive *toxic* wastes mixed with sewage, then unlawfully dumped there, Fine continued, stressing that DV received a permit to dump only *non-toxic* materials at the farm; DV was operated by Vincent DeVito, Tommy Milo's partner in the Al Turi landfill. Alleged Fine, "Vincent DeVito has been heavily involved in organized criminal activities, including loan sharking. Thomas Milo is a Genovese-Tieri associate." All County tank trucks and other tankers allegedly dumped loads at the sod farm, and that material dropped into a canal leading to the Warkill River, Fine revealed.

He further named Mayor John A. Lynch of New Brunswick, N.J., as a representative of a toxic waste disposal company Fine alleged was tied to the mob. Fine charged that the Lynch company, A to Z Chemical Co., had some of the same owners as two other companies at the same address with ties to mobsters, Jersey Sanitation and J&B Disposal. Naming owners, Fine alleged, "J&B and these principals have been linked to organized crime." Among them, he said, was George Katz, a New Jersey political fund-raiser and hauling executive who became a witness in the Justice Department's Abscam investigation into political corruption. He said Katz was "a partner and business colleague of certain persons associated with organized crime."

Fine claimed some segments of the industry were "dominated by hoodlums associated with organized crime," and attacked government officials for failing to protect the public from being poisoned. He charged that, "In some cases, those officials have been involved in decisions that allowed them to be poisoned." In October, 1979, Fine alleged, investigators raided a Staten Island dump where toxic chemicals were oozing out of 10,000-gallon storage tanks, eating away at the barrels, and spilling into waters of Arthur Kill. Fine said the site had been leased to Positive Chemical Co., which, he charged, hadn't been certified to do business in the state. Fine said the company later changed its name to Chelsea Terminal. Toxic waste, Fine went on, was dumped there by Samson Tank Cleaning Co. trucks using bins of Jersey Sanitation. He said Jersey Sanitation was owned by Katz, Patsy and Frank Stamato, and others. Frank Stamato, he alleged, "was linked to Gerardo Catena," who, he asserted, is a Cosa Nostra lord. "Jersey Sanitation is also J&B Disposal," Fine contended, and officers and directors of J&B are John Albert, Katz, Eugene Conlon, and Patsy and Frank Stamato. Fine added that Albert and Conlon—who were also involved with Chemical Control—are also directors "of the notorious toxic waste polluters, A to Z Chemical"; moreover, he charged, Mayor Lynch was "the registered agent" of A to Z, and calls to A to Z were referred to either J&B or Jersey Sanitation.

Fine was fired from his job heading the New York office of the State Organized Crime Task Force. He believes his firing resulted from his crusade against "corruption in criminal investigations" of poisonous waste dumping.

Where have all the accusations led? In April, 1983, two New Jersey waste haulers and their allegedly mob-tied companies were found guilty. They are Louis Spiegel and Anthony Scioscia and their respective firms, Inter County Refuse Service and Home and Industrial Disposal Service; however, cases against John M. Gentempo and Louis Mongelli and their respective firms, Browning-Ferris Industries and ISA, ended in mistrials. Furthermore, the judge strictly banned any statements referring to mob links at the trial.

Florida has also been invaded by Big Apple mob-connected garbage haulers. The U.S. Justice Department's Organized Crime Strike Force in Tampa tried a criminal racketeering case against 11 people, five of whom were also named in a Florida civil anti-trust suit. The five were alleged by the FBI to be Mafia members or associates from New York. They were charged with conspiring to seize control of garbage disposal in Florida by "getting tough." The criminal indictment cited threats, extortion, pistol-whippings, beatings, and the burning of garbage trucks belonging to Florida trash haulers who resisted Mafia takeover. The 11 mob-associated defendants included Tampa Mafia lord Santo Trafficante, Jr. According to Robert Merkle, U.S. Attorney in Tampa, in a Feb. 1, 1985, statement, "All 11 were found guilty and given jail terms in the criminal racketeering case." Moreover, in the civil anti-trust matter, according to Larry Evans, attorney with the anti-trust section of the Florida Attorney General's office in Tallahassee, "The anti-trust case did not go to trial since it was settled before trial with most defendants being assessed civil penalties totaling $56,000, and the rest of them receiving default judgments totaling $26,500 against them."

Cracking Down

In 1980, Congress passed the Resource Conservation and Recovery Act. This new law makes it a felony to purposely break poisonous waste disposal guidelines. Federal jurisdiction is crucial, since dumpers often cross state lines.

As of April, 1983, the U.S. Justice Department's new environmental crimes unit had 25 grand juries in cities from coast to coast impaneled to look into possible felony indictments of individuals and companies who knowingly broke laws in getting rid of hazardous chemicals. Most of these cases were brought about through the Resource Conservation and Recovery Act and involved grand juries in New Jersey, New Hampshire, Pennsylvania, Virginia, West Virginia, Michigan, Ohio, Florida, Kentucky, Tennessee, Illinois, Missouri, Nebraska, Texas, New Mexico, Idaho, and Oregon.

In a case in Concord, N.H., four executives and their Massachusetts tannery, A. C. Lawrence Leather Co., were convicted of storing and disposing of cancer-inducing chemicals in an illegal manner. A Houston, Tex., case even had international repercussions. There, Charles W. Nugent and Ivan Matula were put on trial for supposedly ditching PCB's and other dangerous wastes across the border in Mexico. According to government prosecutors, Mexican citizens subsequently got sick when they employed those barrels to hold drinking water, not realizing they had contained PCB contamination.

On the state level, New York, Connecticut, and New Jersey passed similar laws; however, authorities have neither resources nor experience to detect violations and enforce these new laws. This inadequacy is illustrated by New York State's environmental agency, which made a gesture of sorts toward enforcement by putting a number of conservation officers through a two-week training course! In Washington, the EPA finally organized a team of investigators—but there are only 25 of them for the entire U.S. This is unfortunate, since tough enforcement could make a difference, as could bad publicity for midnight dumpers' corporate clients.

Federal and state governments must also make an effort to get mobsters out of this business. Although mob involvement is heaviest in New York and New Jersey, it seems to be spreading elsewhere.

Is it pure coincidence that there are puzzling cancer clusters in Rutherford, N.J.? There were six cases of leukemia in that city, and statistics show that even one would have been too many to be a chance occurrence. In Rutherford's northeast section, there were also nine cases of Hodgkin's disease (a form of cancer). Why remains a medical mystery, but there are 42 chemical plants within three miles of the town!

Think about it. At this very moment, down some dark, lonely back road in *your* town, some crime lord or lawbreaker not linked to the mob may be rolling his truck in to dump poisons that will ooze into land and rivers, poisoning our children. We don't need the Russians to destroy us; we are doing a dandy job by ourselves.

RISKY BUSINESS[3]

The one thing a big company caught burning toxic waste is not entitled to is surprise at the furor that ensues when the fact becomes known. It happened recently, for example, to no less than General Motors. GM was accused of burning thousands of gallons of cancer-producing toxic wastes at its Livonia manufacturing plant in suburban Detroit, possibly spewing into the air deadly PCBs, dioxins and other chemicals of that ilk. The incident set off alarums at both Michigan and Ontario environmental agencies, generating the kind of press that makes for public relations nightmares.

It was, as it happens, no fault of GM's. It had bought 500,000 gallons of boiler fuel from Can-Flow Services across the river in Sarnia, Ont., and by accident or design Can-Flow had mixed anywhere between 50,000 and 200,000 gallons of toxic waste into the brew. But it's just the sort of accident that has made hazardous waste one of the hottest political issues in the country these days. It was the central issue in last month's gubernatorial contest in New Jersey and in local campaigns beyond counting. The issue has torn Congress into bitter factions over the extension of the five-year-old Superfund for cleaning up environmental hazards.

[3]Reprint of an article by James Cook, *Forbes* staffwriter. Reprinted by permission from *Forbes*, 136:106–07, 109, 112, 115, 118, 122. D. 2, '85. Copyright © 1985 by *Forbes*.

Politically, it's an issue with only one side. Nobody in Congress, or in New Jersey, is indifferent to the problem of hazardous waste. Nor is there any real confusion about what these wastes are or the hazards connected with them. They are substances, over 450 at last count, mainly the by-products of industrial processes, that pose a threat to public health and safety—most commonly because they are linked with cancer and similar life-threatening ailments. But there is controversy all the same. It's over what should be done about hazardous wastes and how much money should be committed to doing it.

The problem comes not just from wastes currently being generated but from those improperly disposed of since the industrial revolution began more than a century ago. Such wastes range from lead to mercury, from benzene to dioxin, and their danger comes mainly from improper disposal—by imperfect (and illegal) incineration (as at GM's Livonia plant) that contaminates the air or by improper storage in ponds, wells or landfills that can comtaminate groundwater.

These hazards are anything but academic. In West Milford, N.J., to recall one notorious example, the local teamsters representative offered the town some waste oil to tame the dust on its backcountry roads. Free of charge. The oil turned out to be laced with highly toxic PCBs. In saving hundreds of dollars on oil, the town wound up spending thousands to dig up a 1.5-mile stretch of road it had applied the oil to. And West Milford was lucky. Residents of Times Beach, Mo., after a similar experience, last year had to abandon their town.

West Milford's benefactor was Bayonne's mob-connected Sampson Tank Cleaning Co. But how do you explain Times Beach? Its supplier was a legitimate local businessman who didn't know what he was doing. Elsewhere the problem is more than that. Says the Environmental Protection Agency's director of enforcement, Thomas Gallagher: "It costs between $100 and $200 to dispose of a 55-gallon drum legally. If a generator or a crooked transporter will do it for $50, everybody makes money."

Such environmental opportunism is not limited to the mob or to amateurs. Rollins Environmental Services once mixed some hazardous waste with crude oil and shipped it to an Ashland Oil refinery in Kentucky. The refinery was so badly damaged that Ashland sued. Browning-Ferris Industries once mixed nitrobenzene with oil and gave away the resulting brew as a surfacing agent. Former Texas congressman Bob Eckhardt once calculated that improper disposal could have turned the estimated 13 cents a gallon that BFI might have lost disposing of the waste in a landfill into a possible 31-to-49-cents-a-gallon profit. The EPA got after Waste Management last winter alleging, among other things, that the company saved $20 million in incineration costs by mixing PCBs with recycled oil and selling it in the Midwest for heating and road-oiling purposes. Waste Management, making no admissions of guilt or innocence, paid a $2.5 million fine and signed a consent decree.

Are such examples mere anomalies? Eckhardt, for one, has had his doubts. "In view of the enormous economic incentive for improper disposal of these toxic materials," he says, "only a strong corporate ethic pervading all levels of management or a very effective regulatory system could prevent such materials from escaping into the environment and threatening the public health."

Until recently, at least, the U.S. hazardous waste industry had neither. Indeed, until 1976, when Congress enacted the Resource Conservation & Recovery Act (RCRA), there were no rules at all. What has happened since then in the regulation of hazardous waste amounts to an economic as well as an environmental revolution. What had always been a commodity business—solid waste disposal—became a specialty business overnight. "When we did the Ford plant," one Harold Kaufman, a former garbageman turned FBI informant, recalls, "the sludge was going for $2 a yard. With the stroke of a pen, we raised the price to between $35 and $85 a yard. We commingled the sludge with shale from Chevron, and the New Jersey [Department of Environmental Protection] allowed us to take this to a regular landfill, which we did for $3.75 a yard. In less than six months, the company made $600,000." Hazardous waste was not just a new business. It was a bonanza.

The industries that generate hazardous waste in the U.S.—primarily chemicals, mining and petroleum—usually dump it in pits or mines or landfills, hold it in ponds or lagoons or storage tanks or, less frequently, dispose of it in public landfills. And, up to a point, quite properly. Waste can best be disposed of at the point of production, before it gets commingled with other substances, so that all but 4% of the estimated 265 million tons of waste generated each year is stored or disposed of on site. The rest is treated or disposed of mainly in commercial landfills or incinerators by subsidiaries of Waste Management (which claims 40% of the business), Browning-Ferris Industries and Genstar, or by independents like Rollins Environmental Services, International Technology and Environmental Systems.

Altogether, the business runs around $5 billion a year, of which only a quarter goes to the commercial operators. But the business is growing fast—from $1.2 billion in 1980 to $3.5 billion in 1984. A study by the Congressional Budget Office figures it could easily reach $10 billion by 1990, with the bulk of the gain going to the commercial operators.

The market is growing more because of increased regulation than increased volume. Whereas EPA originally established strict disposal requirements for some 300 different substances, it now regulates over 450. And where originally only large waste generators were regulated, a year ago Congress put most smaller waste generators—dry cleaning establishments, gasoline stations—under regulation as well, and so enlarged the commercial market by close to 8%. As regulation grows increasingly stringent, waste generators of all sizes are more often turning to commercial operators to do the job.

This is no mom-and-pop affair, as garbage collection often is. It's a highly specialized business requiring skills, technical sophistication and, most of all, capital. Landfills cost $5 million or more to get into operation, and high-temperature incinerators can cost $30 million.

It's also a national business—unlike solid waste—and a service valued highly enough to justify shipping waste generated in Massachusetts to a landfill in Alabama, or from New York to an incinerator in Chicago. The big waste management companies control the bulk of the commercial disposal sites—landfills, deep wells or incinerators. Thousands of broker-haulers, some legitimate, some not, collect the waste and dispose of it—legitimately if they can, illegitimately if they have to.

Though environmental experts once estimated substantial quantities were dumped illegally, nobody really knew. Nobody knows now. EPA thinks it's less than 10%, but whatever the percentage, it is almost certainly on the rise, as small waste generators now coming under regulation for the first time find the economic penalty of proper disposal burdensome. "The incentives for illegal disposal," says EPA's Gallagher, "are going to increase geometrically."

The financial rewards are now so tempting that even the mob has plenty of independent competition—from sleazy businessmen who don't scruple to make an environmentally unsound dollar, from once-legitimate businessmen who would rather not know how the waste is disposed of, from underlings in the big corporations who score brownie points by keeping their disposal costs down.

Environmental protection agencies on both the state and federal level have set up elaborate tracking procedures to assure that hazardous wastes are disposed of properly, but circumventing them is easy. You just falsify the manifest. That's how Ciba-Geigy, indicted in New Jersey in October, was charged with having dumped toxic wastes illegally for more than a decade. And, until EPA began tightening the rules last year, once you crossed territorial jurisdictions, anything went. Says Jeremiah B. McKenna, chief counsel to New York's Select Committee on Crime: "Canadian waste is dumped in the U.S. The EPA stops looking at the border. You take New York waste and dump it in New Jersey. Take Jersey and dump it in New York. Nobody tracks it. You have left state jurisdiction."

Whether unscrupulous or actually mob-dominated, outfits that make a business of improperly disposing of toxic wastes have their own rules. "It's not in organized crime's interest," says Ronald Goldstock, head of the New York State Organized Crime Task Force, "to mix the toxic in with the solid waste. It's too big a loss to take, if you dumped toxic waste in a landfill and lost it [through government action]. They're going to dispose of it in other ways—mix it with oil or cart it outside their traditional areas."

It's now illegal everywhere in the U.S. to mix waste with recycled oil, but that isn't enough to stop the business. In New Jersey members of the Grungo family will be retried on charges of selling contaminated No. 4 oil in five states, from Maryland to Massachusetts, to heat apartments, office buildings, industries, schools and churches.

"All you have to do is stand on the George Washington or Verrazano bridge and you'll see these trucks lumbering across," says McKenna. "We called up one fuel oil dealer at the height of the season, and they were offering us fuel oil at substantially below market. The beauty is they are getting paid twice—to take away the toxics and also to provide the oil."

All too often, everybody's in it together—sleazy operators and great corporations alike. Take Russell Mahler, a Connecticut millionaire. Through 11 companies, Mahler handled hazardous waste for at least 14 major corporations, blue-chip customers like Chrysler, United Technologies, Public Service Electric & Gas and Ingersoll-Rand.

Mahler had no adequate disposal facilities of his own, so he bribed the superintendent of New York City's landfills, John Cassiliano, to let him dump millions of gallons of hazardous wastes in with the garbage, polluting the landfill and contaminating the groundwater that ran underneath it. Eventually Mahler was caught and went to jail. So did Cassiliano.

But the responsibility didn't end there. Confronted with millions of dollars in cleanup costs, New York City has now indicted 14 of Mahler's corporate customers, charging that they knew, or should have known, that Mahler was disposing of their wastes improperly. "According to Mahler, one company paid 10 cents a gallon to dispose of a particular waste, " says McKenna, "but it costs $1 a gallon to neutralize it before you can even begin to dispose of it."

The cost of disposing of new waste is as nothing beside the cost of cleaning up those thousands of hazardous waste sites that threaten the public health and safety, some built up over a half-century or more. Not just those hellholes that have entered the mythology of late-20th-century environmentalism—New York's Love Canal, Kentucky's Valley of the Drums, California's Stringfellow Acid Pits—but 60 or more solid waste landfills where garbage and toxics were commingled over the years. It's a massive undertaking, with 541 sites on the EPA priority list alone and as many as 2,000 likely to need urgent attention.

They are the reason Congress created the $1.6 billion Superfund five years ago—financed largely by taxes on petroleum and feedstock chemicals and designed to provide the front money for cleaning up spills and other emergency situations. The first major Superfund project, for instance, was launched after the 1981 explosion at General Disposal's drum storage site in Los Angeles. Superfund laid out $1.4 million to get the cleanup started, and Inmont Corp. assumed responsibility for the $700,000 needed to complete it.

The U.S. surely needed some mechanism for financing cleanup situations where companies were insolvent, bankrupt or defunct. The Stringfellow Quarry Corp. is a case in point. It went bankrupt in 1974 after 33.9 million gallons of hazardous wastes were discharged into the Stringfellow Acid Pits—unlined evaporation ponds—by over 200 parties. Bankrupt, too, is New Jersey's Lone Pine Landfill, which continues to leach toxic chemicals into the headwaters of the Manasquan River, where the state was developing a $40 million reservoir project. Lone Pine had only $100,000 in assets after it was closed in 1979, not enough even to pay the cost of covering the waste with fill. And it still isn't cleaned up.

Superfund so far has been more sound than fury, perhaps because EPA officials like former chief Anne Burford deliberately tried to block the program. Only ten sites have been cleaned up so far, but cleanups at hundreds more are currently in process, and the Administration, Senate and House are wrangling now over how big a renewed Superfund should be—$5 billion, $7.5 billion, $10 billion. Even $10 billion is unlikely to be enough. The Department of Defense estimates it will cost between $5 billion and $10 billion to clean up some 473 military sites alone. The General Accounting Office puts the total national cleanup cost at $39 billion. The Office of Technology Assessment thinks it's well over $100 billion.

All too often, as the OTA study points out, Superfund money has been devoted to moving wastes from one problem site to another, from Dartmouth, Mass., say, to landfills in New York and Ohio that were themselves possible problems. Some Love Canal wastes went to BFI's Cecos landfill only 5 miles away. Waste Management was hired to handle the $7.7 million cleanup of the Seymour, Ind. site, which includes some Waste Management waste. And even the sites to which the waste is moved may not be secure. "The requirement that no unit be leaking would probably rule out all the operating landfills in the country," states William Myers, a former EPA scientist.

"Superfund was designed for emergency situations like Love Canal," says Hugh Kaufman of EPA's Hazardous Site Control division, "and the idea was that people who created the environmental and public health emergencies would have to pay ultimately. What Superfund has been turned into by the politicians is a public works program, and they've done this at the urging of companies like Waste Management, which not only create these catastrophes but also clean them up and get paid for it. Of the $1.6 billion committed in the first five years, we have returned only $21 million to the fund."

The dimensions of the problem explode even as the means of solving it dwindle. Last year Congress limited or barred use of land disposal for certain substances and decreed that all forms of land disposal should be barred within 66 months. But good intentions are no substitute for hard thought. As it is, 903 storage and treatment facilities, 42 incineration facilities and 186 land disposal sites have been closed. That leaves, at last count, only 373 commercial treatment facilities, including perhaps 36 commercial incinerators and 58 commercial land disposal sites in all the U.S. Clearly inadequate to the need.

But if not land disposal, what? There are several alternatives. Some hazardous wastes can be recycled, others transformed into chemically stabilized solids, and still others stored in secure, deep wells or landfills. Other wastes can be treated chemically, biologically or thermally—that is, burned—and safely disposed of in a nontoxic landfill at $15 to $30 a ton rather than $100 or $200.

For those wastes that will burn, about 25% of the total, incineration offers the most promising solution, though hardly a cheap one. That's why both Waste Management and At-Sea Incineration, an affiliate of bankrupt Tacoma Boatbuilding, are eager to begin burning toxics offshore. Ocean incineration is significantly lower cost, but it has been opposed by both Rollins and Environmental Services, which together operate two of the four land-based, PCB-approved incinerators. The U.S. will need all the incinerators it can get. EPA figures it will need 82 land-based and 33 ocean-going incinerators by 1990.

"The next few years look very bright, but it will be very expensive to get service," Eugene Wingerter, head of the National Solid Wastes Management Association, says cheerfully. "This industry thrives on regulation."

Law enforcement people worry that as more and more landfills close down, toxic wastes like PCBs and dioxins will end up in the high-temperature waste recovery plants that are springing up all over the country, where they may or may not be incinerated properly. Such concerns are by no means academic. Hazardous waste in the garbage stream has already destroyed a waste recovery plant in Akron, Ohio.

Incinerators aside, it has become increasingly difficult to get a hazardous waste project of any kind into operation. Nobody wants a hazardous waste site in his own backyard, and states with disposal facilities are beginning to balk at becoming dumping grounds for states without them. The capital costs of disposal, treatment and storage have gone through the roof, and the regulatory process is so involved it may take five years to get a project operational. Even worse, most companies can no longer buy liability insurance.

"When you have to have something nobody wants, and when it has major financial liabilities, who do you look to to take on those liabilities? The taxpayers of the U.S.," says EPA's Hugh Kaufman. "These guys at the waste management companies really know how to play the system and make sure the liability does not remain on their shoulders."

It's probably better that the taxpayer assume these costs if the alternative is that no one assumes them. Bluntly, and in short, environmental safety may be too vital a matter to leave to the unfettered working of the profit motive. There is a parallel in the nuclear power business for the past 30 years. Government has assured certain risks in the nuclear business, limiting the utility's liability in the event of an accident and agreeing to assume responsibility for future storage of wastes. The hazardous waste industry can make arguments every bit as compelling for such government support.

It's a risky business, all right, extraordinarily risky for everybody involved—generators, disposers and the public alike. But the magnitude of the opportunity is suggested by the fact that the rewards still seem to outweigh the risks. If the new waste is going to be disposed of, if the old waste is going to be cleaned up, it's the commercial companies, the risks notwithstanding, that will do the job. "There is a lot more upside potential than there is downside risk," says Waste Management President Phillip B. Rooney. "That is what we are counting on. We continue to believe the potential is enormous."

THE EXPORT OF U.S. TOXIC WASTES[4]

Ken Kazarian, president of a waste recycling firm near Los
Angeles, recently received a letter that began:

Dear Sir, aware of your serious problem in the disposal of your toxic and
hazardous waste . . . we have acquired an island in the Philippines suitable as a dump site.

The company that sent Kazarian the letter, Gamma Trading,
lists for its address a private residence in Hillsborough, California, a wealthy suburb of San Francisco. Gamma's president is a
27-year-old Filipino woman named Leana Carlos, whose parents
have been the object of a number of civil lawsuits alleging breach
of contract, of a Federal lawsuit for violating banking laws, and
of a lien for unpaid back taxes. The family's lawsuits and business
transactions are handled by an attorney who has managed the accounts of prominent Filipinos with ties to deposed President Fer-
dinand Marcos. In addition, the family's real estate affairs are
monitored by a bank that has been tied to Marcos in reports by
the *San Jose Mercury News*. Until March 1986 Gamma was acting
as broker for L.P.T., a company that planned to build a multimillion-dollar incinerator for U.S. wastes in the South Pacific. The
Carloses failed to return repeated calls to inquire about Gamma's
operations.

Kazarian placed Gamma's letter in his "nut file," along with
a proposal he had received for turning his company's landfill into
a gorilla habitat. But he knew that the letter, like the problem of
hazardous and toxic waste disposal, was no joke. Gamma Trading
appears to be part of a growing shadow industry that is exporting
those wastes from the United States, particularly to countries in
the Third World. And U.S. government agencies that are supposed to enforce the laws regulating such shipments lack the
money and the personnel to do the job.

The material that is being exported includes heavy metal residues and chemical-contaminated wastes, pharmaceutical refuse,
and municipal sewage sludge and incinerator ash. The risks in-

[4]Reprint of an article by Andrew Porterfield and David Weir, journalists at the
Center for Investigative Reporting, San Francisco. Reprinted by permission from
Nation, 245:325, 341–44. O. 3, '87. Copyright © 1987 by *Nation*.

volved for countries that accept the wastes range from contamination of groundwater and crops to birth defects and cancer.

Traditionally, the majority of U.S. toxic waste exports have gone to Canada, where regulations are less stringent than in the United States. Under an agreement between the two countries, massive amounts of toxic wastes are exported to a landfill site in Quebec Province and an incinerator in Ontario. But the most abrupt increase is in shipments to the Third World, where regulations are either nonexistent or sketchily enforced. The crucial difference, however, is that waste exports to Canada are legal and aboveboard. The rising tide of exports to the Third World cannot easily be documented, for it is frequently concealed. However, using court records, interviews and documents obtained under the Freedom of Information Act, we have been able to piece together a number of previously unreported examples of the shadowy trade in hazardous wastes. Some have come to light because the companies concerned were convicted of illegal activity; others because we managed to obtain export authorization records from the Environmental Protection Agency; and still others because records show that the foreign government involved has refused to accept the wastes.

The sources also demonstrate the degree to which U.S. government agencies are involved in the hazardous waste export trade. U.S. officials who are aware of the sensitive legal and foreign policy questions involved seem reluctant to crack down on illegal dumpers, although recently some of the most blatant offenders have been prosecuted. But the government itself is responsible for generating a significant portion of the hazardous waste exports. One large illegal operation broken up by law enforcement officials last year was obtaining more than half its toxic wastes from various branches of the Federal government, mainly the military.

Underlying the search for new overseas markets is an explosion in the volume of recorded hazardous wastes being produced in the United States. According to the General Accounting Office, the amount rose from about 9 million metric tons in 1970 to at least 247 million in 1984. Other experts place the current figure close to 400 million metric tons—or ten pounds per person per day. Western Europe and Japan produce much less per capita, apparently because of more efficient production processes and a greater awareness of the limits of landfill disposal space. All of

Western Europe, in fact, generates only an estimated 30 million to 40 million metric tons of hazardous wastes annually. The cost of disposing of U.S. wastes has also risen dramatically. In 1976, disposal cost $10 a ton; today the figure is between $60 and $140 a ton, in some cases even higher.

As scientists discover the environmental and public-health problems created by past disposal practices, landfills across the United States have been forced to close. Major concerns include the contamination of groundwater by hazardous chemicals that leach from dump sites and the creation of dangerous breakdown products, such as highly toxic dioxins, during incineration. Within a decade, according to the E.P.A., more than half the states will have completely exhausted their landfill capacity and be unable to accept hazardous wastes, adding further to the cost of transport and disposal.

Since 1980, companies that intend to export hazardous wastes have been required by the Resource Conservation Recovery Act to send notices to the E.P.A.'s Office of International Activities, and since last November they have had to show they have the permission of the receiving country. The number of shipments thus documented has risen from just thirty in 1980 to more than 400 last year. (More than half of those went to Canada.) Although the number of notices for shipments to Third World countries has jumped sharply, from four in 1984 to nineteen in 1986, those figures do not tell the whole story. They do not include, for example, wastes not yet officially classified as hazardous; according to a recent study by the G.A.O., the "E.P.A. does not know whether it is controlling 90 percent of the existing waste or 10 percent. Likewise it does not know if it is controlling the wastes that are most hazardous." There is also the problem of overlapping jurisdictions within the E.P.A. Export of certain chemicals, like polychlorinated biphenals (PCBs), is forbidden by the Toxic Substances Control Act and is regulated by the E.P.A.'s toxic substances division, which does not handle wastes.

Even some of the exports that appear to fall under the E.P.A.'s purview are escaping detection. According to one knowledgeable E.P.A. official, who did not wish to be named, "Many exporters don't bother to give notice to the agency because there isn't any enforcement." Gary Steakley, an E.P.A. enforcement official, added, "We have been promised funding increases for the last few years for enforcement of toxic wastes but have received nothing. Regulations [don't stop] supplies of illegal drugs. Why should they stop dumping of toxic wastes?" By comparing Customs Service records at various U.S. ports with the notices sent to the E.P.A., says one well-placed source, "E.P.A. auditors have detected many more shipments than were reported to the agency. The amount of trade is as much as eight times more. That's not including smugglers who elude customs. I think many countries are having it sneaked in." (This is sometimes with the complicity of officials in those countries.)

While the E.P.A. attempts, within severe budgetary constraints, to regulate the trade in hazardous wastes, other U.S. government agencies appear to be encouraging it, at least to the extent of selling wastes to exporters. In two cases such sales have come to light as a result of criminal proceedings against waste handlers. Until last year, when a Federal court in New York sentenced them to thirteen years in jail for fraudulent business practices, Jack and Charles Colbert were making big money as toxic waste exporters. They had amassed huge volumes of explosive and poisonous wastes in twenty warehouses over primarily the eastern half of the country, from Texas to South Carolina to the Canadian border. Many shipments were sent overseas, to India, South Korea, Nigeria and Zimbabwe. Federal and New Jersey State prosecutors were eventually able to stop the Colberts because they falsely labeled their waste products as pure chemicals, to mislead their overseas buyers. In the case that sent them to jail, the Colberts sold wastes, labeled as dry-cleaning solvent, to a company in Zimbabwe sponsored by the U.S. Agency for International Development.

The Colberts purchased their wastes from the Navy, the Army, the Defense Department and the Agriculture Department. They also bought lead-tainted engraving paper from the Treasury Department's Bureau of Engraving. Fortunately, the paper was never sold. Just after New Jersey authorities seized one of the Colberts' warehouses in Newark, "Some guy came by who wanted to buy the paper for an African country," says Bruce Comfort, an investigator for the New Jersey Department of Environmental Protection. "He was going to have it sold to Africa as toilet paper. He didn't know it was contaminated."

The Colberts' scheme wasn't an isolated case. In 1984, American Electric, a Jacksonville, Florida, company run by businessman Maxwell Cobb, tried to sell PCB-laden wastes to Honduras.

Much of Cobb's material was acquired under a large contract with the Defense Department to handle wastes from military bases up and down the East Coast. According to Bob Yerkes, an Assistant U.S. Attorney who helped prosecute Cobb, "The problem was that there was not enough evidence. The government records were not precise. A jury acquitted him." (Cobb went to jail anyway, on a drug conviction.)

Major U.S. cities, as well as Federal agencies, have joined the waste export bandwagon, sometimes with the approval of the State Department. Among them is Philadelphia, which has faced a rise in waste disposal costs from $20 to $90 a ton since 1980. The city intends to transport as much as 600,000 tons of ash residue a year from its municipal incinerator to Panama, which plans to use the material as landfill for roadbeds. The deal will cost Philadelphia taxpayers only $37.25 a ton. The city has also been trying to find a Third World destination for its municipal sewage sludge, which has historically had a high metals content and has been the subject of a lawsuit against the city by the E.P.A. and environmentalists for violating Federal standards.

Honduras, the first country approached by Philadelphia, rejected the sludge proposal. The next candidate was Guatemala, which may also turn down the waste despite the initial enthusiasm of President Vinicio Cerezo, who saw the deal as a valuable source of foreign exchange, and the approval of the State Department. "We have reviewed the proposal and have no objections," said a memorandum from the U.S. Embassy in Guatemala City to Washington. "One can imagine that shipping sewage sludge . . . will incite some unfavorable press in Guatemala. With this caveat, we have no problem."

Philadelphia's export plans may not go through, but Applied Recovery Technologies, the company planning to ship the wastes, has since approached other cities. U.S. sludge may also end up in the tiny British Caribbean colony of the Turks and Caicos Islands, which proposes to use it as fertilizer. In a letter to Henry Valentino, president of Applied Recovery Technologies, State Department coordinator for Caribbean affairs John Upston wrote, "This project represents a breakthrough in a . . . way to deal with a growing problem for our cities. At the same time it helps in a major way the economy of a small friendly country in the Caribbean."

A differing view was expressed by an official of an overseas environmental organization, who asked not to be named for fear that the A.I.D. would cut off his funding. "I am concerned that if U.S. people think of us as their backyard, they can also think of us as their outhouse. Disposal of wastes on land can have very serious consequences. Toxic materials can be taken up in root systems which can grow in sludge used on land. These materials could end up in fruits and vegetables or in animals."

Nevertheless, in addition to Philadelphia, Boston, Hartford, Los Angeles and Washington have also reportedly been looking into exporting their municipal sludge to the Caribbean and more deals are expected soon.

"The Caribbean is a big target area," says Dr. Noel Brown of the United Nations Environmental Programme in New York. "It's cheaper to barge it down there than to move it overland forty miles. We don't have an international policy on wastes. We need one now."

The key U.S. government officials responsible for monitoring the burgeoning hazardous waste traffic claim they are powerless to stop even a dangerous export if the host country agrees to accept the shipment. "Under the Federal system, we only have control over what's in the country," says Wendy Grieder, an official in the E.P.A.'s Office of International Activities. "Once it leaves, we can't do anything about it. And their destinations don't often regulate as tightly [as does the United States], even Canada. Once it gets there, we don't know what happens to it. That's why we had so many exports."

In addition to the E.P.A., the Commerce Department monitors some hazardous waste exports. Under the 1977 Export Administration Act, the department issues permits for exports of strategic metals that are contained in ash and other factory residues. In the process the department gathers information about the overseas movement of these residues. Since 1980, thirty-five of the seventy-five approved destinations for these exports have been underdeveloped nations, including the Philippines, Mexico, and many in Central America and the Middle East. But the Commerce Department does not monitor what happens to the mountains of waste materials, which may contain lead, mercury and other dangerous substances, after the tiny fractions of usable metals have been extracted. This is partly because the intent of the 1977 law was to restrict the export of strategic goods, not to assess the health and environmental hazards involved. "After [the waste material] gets there, the country can do whatever it wants with it," said a department official who requested anonymity to avoid being fired. "I assume it gets tossed out."

Of course, simply "tossing out" hazardous wastes can cause severe public-health problems. In the Dominican Republic, says one expert, wastes containing antibiotics and fish oil were imported for use as cattle feed and fertilizer from a U.S.-based Abbott Laboratories' facility in Puerto Rico, where pharmaceutical dumps are full. In 1985, one animal died from the mixture. According to Dr. Antonio Thomen, director of the Dominican Institute for Bioconservation, ingestion by humans can cause hormonal disorders, birth defects and severe intestinal illnesses, particularly among children. The Dominican Congress has now passed a law prohibiting the import of pharmaceutical wastes as a human-health hazard.

E.P.A. files also contain records of other shipments that the agency barred from export after the importing countries refused to accept them. In 1985 the Costa Rican government denied entry to a shipment of 205 cylinders of poisonous, corrosive gases after asking E.P.A. officials for information about the materials involved. Eco-Therm, a California transport company hired by T.R.W. Inc. of Redondo Beach, California, to make the shipment then sued several E.P.A. officials for obstructing trade, but the case was dismissed early last year in Federal court in Washington for lack of evidence.

In another incident, when the Bergsoe Metal Corporation of St. Helens, Oregon, tried to send 700 metric tons of crushed battery plates containing lead to the Kwang Shin Industrial Company of South Korea, the Korean government told the State Department in a telegram that it was opposed to the import of "such harmful waste." The E.P.A. rescinded its export approval. Bergsoe then attempted to get permission to ship the waste to three successive companies in Taiwan, but that government also intervened and the E.P.A. could not approve the deal.

In addition to individual shipments of waste products, at least two American companies have proposed multimillion-dollar deals to build incinerators in the South Pacific to process U.S. wastes. L.P.T., a company with offices in American Samoa and California, is seeking approval to build an incinerator in American Samoa to burn U.S. wastes and export the ash to the

Philippines, where it would be used as landfill. The Samoan government has not given its approval according to the E.P.A., and the Philippine government asserts that it has a policy of not accepting hazardous wastes from anyone. However, in April, according to local newspaper reports, a woman named Elvira Medua Patel, who said she was a special envoy of President Corazon Aquino, showed up in American Samoa, saying Aquino wanted the waste shipments to go to the Philippines and was even willing to have the incinerator built there, if American Samoa turned down L.P.T.'s proposal. The Philippine government denies that it offered any such deal.

Also, a company named Western Pacific Waste Repositories, based in Carson City, Nevada, is proposing to build a hazardous waste storage and treatment plant on Erikub atoll, an uninhabited area of the Marshall Islands. Company president Dennis Capalia proposes to use the atoll to store wastes for the next 100 years. In return, it will make cash payments and promote the development of nearby Wotje atoll, whose residents now use Erikub to make copra and hunt for marine life.

In the United States, the pressure to ship wastes overseas is growing. Representative James Florio, who conducted hearings on such exports in 1983, says, "Like water running downhill, hazardous wastes invariably will be disposed of along the path of least resistance and least expense. Conditions are ripe for finding 'safe havens' for hazardous wastes around the globe."

Many experts see the trend as part of an old American habit of putting undesirable elements out of sight and out of mind. "To push things out beyond its borders, that is a traditional American philosophy," notes Nikolai Zaitsev of the United Nations' Center on Transnational Corporations.

Despite the growing problem, the Reagan Administration has shown little concern. Grieder of the E.P.A. defends the agency's "hands off" policy on exports to the underdeveloped parts of the world. "At E.P.A., we're not in a position to say, 'That's a bad deal,' or 'They don't know what they're doing.' If the receiving country says yes, there's nothing we can do about the shipments. Although morally it's probably a good idea, as an agency we can't do it. What if we did stop a shipment? What if the government makes a mistake and the company and the shipment are legitimate?"

But some officials worry about the possibility of a chemical waste disaster occurring as a result of the U.S. waste export boom. "If I were the U.S. Secretary of State," said Senator George Mitchell in 1984, "I would want to be sure that no American ally or trading partner is saddled with U.S. wastes it does not want or does not have the capacity to handle in an environmentally sound manner."

There is even the possibility that exported wastes may return to haunt the United States in a more direct way. Certainly the danger is apparent to Grieder. "It hasn't happened yet, but it could happen," she says. "It's possible that we could send sludge to the Caribbean and they might use it on, say, spinach or other vegetables. We would get it back here, and the F.D.A. would say, 'Hey, wait, you've got too much cadmium in those vegetables.'" Since the Food and Drug Administration checks only a small portion of foods and vegetables that come into the United States, exported hazardous wastes could easily end up on our dinner table.

Meanwhile, overseas pressure to prevent the wholesale dumping of U.S. wastes is building despite U.S. government encouragement and the large sums debt-ridden Third World states can make from such deals. "Governments could fall because of this," says the United Nations' Brown. "It would be worse than sending guns, because it affects everyone."

BIOLOGY'S ANSWER TO TOXIC DUMPS[5]

Rain falls as a rubber-gloved technician shovels mud and hauls it to a nearby lab. For years a Minnesota sawmill dumped its wastes here, and for years rain has been leaching traces of the wood preservative pentachlorophenol (penta) into the groundwater. But now scientists are using the site, one of more than 500 penta pits across the nation, to study the possibility and profitability of using bacteria to clean up hazardous wastes.

The concept is not new. Bacteria have been used to treat municipal wastewater for almost a century. But during the last few

[5]Reprint of an article by James J. Holbrook, free-lance writer. Reprinted by permission from *Sierra*, 72:24–25, 28. Ja./F. '87. Copyright © 1987 by *Sierra*.

years biochemists have been showing that nature can provide organisms capable of metabolizing some of our most troublesome toxins.

The microbe being tested at the Minnesota site is called *Flavo bacterium*. Ron Crawford of the University of Minnesota's Gray Freshwater Biological Institute isolated the bacterium by continuing to feed a sample of penta-contaminated soil a steady diet of penta. "Over a period of several months, any organism that can't metabolize penta will not survive," Crawford explains. *Flavo bacterium* converts penta into harmless carbon dioxide and chlorides.

Another researcher at the institute, John Wood, is experimenting with an alga that can convert dissolved heavy metals such as lead and mercury into insoluble forms, thus keeping them out of the food chain.

At the Woods Hole Oceanographic Institution in Massachusetts, microbiologists Graig Taylor and Holger Jannasch have maintained a colony of sulfur bacteria collected near deep-ocean vents. In their natural habitat these organisms live on the hydrogen sulfide leaked by the vents, so they may be useful in removing sulfides, a primary cause of acid rain, from the emissions of coal-fired power plants.

In another promising development, Michigan State University biochemists Steven D. Aust and John A. Bumpus recently reported that the same enzyme system that enables the fungus *Phanerochaete chrysosporium* to decompose dead trees is also capable of reducing several deadly and persistent poisons—DDT, lindane, PCBs, and dioxin—to carbon dioxide.

These and other lab triumphs have demonstrated the feasibility of using biological methods to clean up hazardous wastes. It's not yet clear, though, whether this technology will succeed in the marketplace.

"What's holding up the use of biological treatments is the market," says Sierra Club Political Director Carl Pope, co-author of *Hazardous Wastes in America* (Sierra Club Books, 1982). "The demand hasn't been there, because it's cheaper just to dump the stuff."

But as dumps grow more controversial and expensive—and biological techniques grow more sophisticated—the market is beginning to respond. In 1985, for example, the Rothschild Ven-

ture Fund and the Plant Resources Venture Fund provided
$500,000 and the promise of significant additional funding to the
newly formed Minnesota firm Biotrol, Inc. The company now offers a penta cleanup service that uses Crawford's *Flavo bacterium*.
Biotrol is one of several U.S. firms trying to apply the ideas of
modern microbiology to toxic waste disposal.

"The hazardous waste industry is on the verge of an
explosion," says Minneapolis attorney and Biotrol co-founder
Lindsay Arthur, "much the way the computer industry was 20
years ago. It's likely to be a multibillion-dollar-a-year industry."

Technical problems remain, however. Biological treatment is
not simply a matter of pouring a slurry of microbes into a pit and
billing the customer. "You've got to provide an environment the
bugs can thrive in," says Boyd Burton, president of Biotrol. "You
need proper levels of oxygen, along with a suitable pH and temperature. Naturally occurring conditions do not support the right
kind of bacteria."

One possible solution involves placing contaminated soil and
water into self-contained "bioreactors" that will allow technicians
to control surrounding conditions. Although it will require significant outlays of capital, this is the approach being explored by
Biotrol.

Another solution involves altering the micro-organisms. "In
the future we will use genetic engineering to modify the bugs, to
give them a greater tolerance for their environment and a more
voracious appetite," Burton predicts.

Environmentalists are hopeful, but approach the industry
with caution. "Even the introduction of naturally occurring organisms into new environments can be disruptive of ecosystems,"
says Pope. "When you go beyond that to genetic engineering, you
have to be very careful you know exactly what you're doing to the
environment."

Burton accepts that responsibility willingly: "There will be
many skirmishes, many questions about whether we know
enough," he says. "It will be our obligation to know enough."

In Burton's opinion, the new technology's success is just a
matter of time. "In a decade or so we will not only be able to clean
up the mess we have now, but we will be able to handle the waste
we produce each year. We're a long way from that today, however. We've just entered the slugging match of development and
demonstration of practical commercial processes."

Environmentalists are keeping a close watch on the emerging industry. If it can fulfill even part of its initial promise, we could soon be treated to the spectacle of watching profit-making companies compete to clean up the Earth.

DIOXIN[6]

Dioxin. The word evokes a variety of reactions. Much of the public worries that the compound will cause poisoning even at minimal exposure. Toxicologists, knowing the severe toxic effects of dioxin in experimental animals but being uncertain about comparable serious effects on people, call for more research. Regulators, who must make decisions based on this conflicting evidence, are left wondering what to do. As one who has spent many years studying the dioxin issue, I hope in this article to provide a useful perspective for making judgments about the potential hazard of the material.

Dioxin is actually a name for a family of chemical compounds. The name refers to their basic structure: two oxygen atoms joining a pair of benzene rings. Substitution of chlorine atoms for hydrogen atoms on the rings produces a chlorinated dioxin, of which there are many. The chlorinated dioxin of interest here is 2,3,7,8-tetrachlorodibenzo-p-dioxin, usually abbreviated to TCDD. It is a by-product of the manufacture of trichlorophenol, which serves in the manufacture of two herbicides (the best-known being 2,4,5-trichlorophenol, or 2,4,5-T, one of the ingredients of Agent Orange) and the antibacterial agent hexachlorophene. TCDD is also produced in a variety of combustion processes.

The legacy of Agent Orange, Times Beach, Seveso, several industrial accidents and other instances of human exposure to significant amounts of TCDD is widespread concern about adverse health effects resulting from minimal exposure to the material. Yet none of the many studies directed at this question have

[6]Reprint of an article by Fred H. Tschirley, *Scientific American* staffwriter. Reprinted by permission from *Scientific American*, 254:29–35. F. '86. Copyright © 1986 by *Scientific American*.

demonstrated that TCDD causes severe chronic human effects. Moreover, not one human death has been attributed to TCDD, even though exposure has been high in a number of cases. The issue points up the broader problem of the difficulty faced by regulators who must make judgments on the basis of incomplete scientific knowledge on the one hand and public fear on the other.

TCDD was first recognized in 1957 as a contaminant of 2,4,5-T, when 31 workers involved in the manufacture of the herbicide in West Germany developed the dermatologic affliction now called chloracne. It is a skin eruption resembling acne and takes its name from the fact that it is caused by exposure to various chlorinated organic chemicals.

General awareness that TCDD is a potential hazard to health and the environment arose in 1970, when a House subcommittee held a hearing on "Effects of 2,4,5-T on Man and the Environment." The hearing dealt among other things with a study by the Bionetics Research Institute showing that 2,4,5-T caused birth defects in animals. Testimony suggested that the teratogenic component of 2,4,5-T may have been TCDD. In the sample of 2,4,5-T tested by Bionetics the dioxin occurred as a contaminant at an extremely high concentration: 27 ± 8 parts per million.

Since then there has been a steady accumulation of information about the sources of TCDD, its environmental fate and its toxic effects. One well-defined source is the formation of TCDD during the manufacture of 2,4,5-T. The amount of TCDD formed increases as the temperature of the reaction and the pH (degree of alkalinity) increase.

In 1977 investigators in the Netherlands reported that polychlorinated dibenzo-p-dioxins (PCDD's) were present in the fly ash from a municipal incinerator. Typically such an incinerator burns, among other things, organic wastes containing chlorine. Similar reports soon came from Switzerland, Canada and Japan. It was believed the compounds resulted from the condensation of chlorophenols. Later quantitative data demonstrated from a wide variety of combustion sources the presence of TCDD's that could not be explained on the basis of preexisting polychlorinated phenols.

These findings led R. R. Bumb of the Dow Chemical Company and 12 of his co-workers to put forward in 1980 the hypothesis that PCDD's can result from trace chemical reactions in fire. The hypothesis has been challenged because the reactions have not been defined. Nevertheless, PCDD's have now been found in the effluent and ash of so many combustion processes that there is no longer serious argument about their formation during combustion, even though the precise nature of the process remains obscure.

Moreover, TCDD has been specifically identified in soil and dust from numerous places, in soot from the chimneys of wood furnaces, as residues in river fishes (some from rivers whose watersheds do not have industrial operations known to form TCDD), as residues in the eggs of herring gulls and recently in adipose tissue from more than 100 people in Canada, the U.S. and Vietnam. It appears that TCDD is a ubiquitous chemical, particularly in industrialized nations.

One might then ask why it was not detected sooner. For one thing, no one looked for it seriously before the 1970's. At the time it would not have been found anyway, except in unusual circumstances, because analytic chemists were only able to detect concentrations of a few parts per million. Since then the detecting equipment has improved at least a millionfold, so that concentrations of a few parts per trillion are now detected routinely. In addition dilution and the destruction of TCDD by light may reduce concentrations to undetectable levels. As analytic technology improves, one can expect that TCDD will be found in many more sites than are currently known.

In places where TCDD is protected from light it is an extremely persistent material. In the early 1970's it was thought the half-life of TCDD (the time required for half of a given amount to be degraded) was about a year; later studies in the U.S. suggested the half-life might be as long as three years. Recent reports from Italy raise the possibility that the half-life of TCDD in soil might be 10 years or even more.

Although precise data are meager, TCDD is known to be strongly held by most soils. The strength of the binding is inferred from the fact that known concentrations of TCDD applied to soil have remained near the surface. Even in the sandy soil examined in Florida the concentration of TCDD in the upper 15 centimeters was as high as 1,500 nanograms per kilogram 10 to 12 years after application.

In the most heavily contaminated zones near Seveso TCDD has been found at a depth of 136 centimeters. Moreover, the concentrations well below the surface were slightly but significantly higher in 1977 than they were in 1976, soon after the accident took place. The presence of soil fissures does not adequately explain this unexpected vertical distribution.

The processes that degrade TCDD in soil are poorly known. Microorganisms do degrade the substance, but at a low rate. S. D. Aust of Michigan State University found a wood-decaying fungus (the white mold *Phanerochaete chryosporium*) that breaks down TCDD without observable mortality of the organism. The rate of degradation is low, but it is conceivable that contaminated sites could be inoculated with the mold to speed up the degradation process.

Sunlight degrades TCDD rapidly by splitting off the chlorine atoms. The reaction requires a hydrogen donor, which is usually available in water or in the wax on leaves. Experiments by Donald G. Crosby of the University of California at Davis showed that 40 percent of the TCDD layered on a glass plate remained after six hours in sunlight; the amount was from 25 percent to negligible when the material was applied as drops on leaves of the rubber plant, but on a loam soil the figure was 85 percent.

TCDD is a highly toxic chemical in experimental animals. The first animal-toxicity test performed is usually a determination of the LD_{50}: the dose that kills half of a test population. Between 1973 and 1978 the LD_{50} of TCDD was determined for eight species. The guinea pig was by far the most sensitive species tested: its LD_{50} for an oral dose was .6 microgram per kilogram of body weight. The hamster was the least sensitive animal tested: its oral-dose LD_{50} was about 1,900 times as high as the guinea pig's. Its intraperitoneal level, 3,000 micrograms per kilogram, was about 5,000 times as high as the oral dose for the guinea pig.

The reasons for the extreme range of acute (short-term) toxicity may relate to the relative speed of clearance from the body. In the hamster half of a dose is removed within 15 days, compared with 30 days for the other species tested. Even though the hamster is much less sensitive than the guinea pig, an intraperitoneal LD_{50} of only 3,000 micrograms per kilogram signifies an extremely toxic material: the toxicity is comparable to that of the insecticide parathion.

Many essentially acute symptoms have been observed in human beings. They include chloracne, digestive disorders, effects on some essential enzyme systems, aches and pains of muscles and joints, effects on the nervous system and psychiatric effects. These symptoms have been transitory except for a few severe cases of chloracne.

Additional tests have measured the chronic, or long-term, effects of TCDD in rodents, rabbits and nonhuman primates. Chloracne, the most sensitive indicator of human exposure to TCDD, has also appeared in rabbits, nonhuman primates and hairless mice. The skin and body can become dry and scaly. A few species lose hair. Some nonhuman primates lose fingernails and toenails without apparent evidence of pain.

TCDD also causes reproductive effects in experimental animals. Cleft palate and abnormalities of the kidney were caused in the offspring of exposed mice at dosage levels of from one nanogram to three nanograms per kilogram of body weight per day. Similar doses in rats caused the death of fetuses. In monkeys a dose of 1.7 nanograms per kilogram per day for two years caused abortions in four out of seven pregnancies.

TCDD is a proved carcinogen in rats and mice. Although the test results vary somewhat, there is fairly good agreement among tests by different investigators. The liver is the primary target in both rats and mice, although the brain, the respiratory system and the thyroid gland have also been involved in a few studies. It is important to recognize that an oncogenic response was reported only after the animal had ingested high doses of TCDD over a long period of time. Rats showed no oncogenic response at dosage levels of from one nanogram to 1.4 nanograms per kilogram per day; mice showed none at dosage levels ranging from one nanogram to 30 nanograms. Moreover, a study by R. J. Kociba of Dow Chemical and his co-workers showed that rats tolerated a daily dose of one nanogram per kilogram per day for two years without showing toxicologic effects.

Nevertheless, the findings from the tests on animals intensified concern about the effects of TCDD on people. On several occasions people have been exposed to "high" levels of TCDD. In this context "high" is a relative term because with one exception—a group of prisoners who volunteered for tests with TCDD—the amount of the material to which a person was exposed is not accurately known. The criterion I employ here to distinguish high from low exposure is whether or not the exposure results in chloracne.

The case of the prisoners is important because known amounts of TCDD were applied to their skin. In the first experiment 60 volunteers were treated with concentrations ranging from 200 to 8,000 nanograms (from three to 114 nanograms per kilogram for a 70-kilogram person), and the dose was repeated two weeks later. The dosages chosen were those that had caused chloracne when they were applied to the ears of rabbits. None of the volunteers developed chloracne, and no other symptoms were observed. The second experiment involved 10 volunteer prisoners who were treated with 107,000 nanograms of TCDD per kilogram. Eight of them developed chloracne, but no other symptoms were noted.

From these experiments one can conclude only that TCDD does cause chloracne in humans when the dose is sufficiently high but that people are less sensitive than rabbits. The tests did not identify a threshold for the development of chloracne in human beings—a piece of information that would be of great value.

The number of people who have been exposed to high levels of TCDD cannot be determined accurately, but it must be in the thousands. Alistair Hay of the University of Leeds has estimated that in the chemical industry alone about 2,000 workers have had high exposure. Low levels of exposure have undoubtedly been experienced by people who handle the herbicides 2,4,5-T and Silvex, in which TCDD was a contaminant; by Vietnam veterans exposed to Agent Orange (50 percent of which was 2,4,5-T); by residents of Times Beach, Mo., where waste oil that contained TCDD was spread on the ground in several places; by chemical-industry workers making the products that include the material, and by many thousands of people who have eaten food (notably fish) containing trace amounts of TCDD or have been exposed to fallout from combustion processes that form TCDD. The total number of individuals with such low exposures probably runs well into the millions.

The possibility of chronic effects from exposure to TCDD causes far greater public concern than that of acute effects. An aspect of this problem about which little is known is the effect of protracted exposure to low levels of the material, as might occur in an occupational setting or from incidental exposures to, for example, the fallout of combustion effluents or to fish that contain low levels of TCDD.

A look at some of the major exposures to TCDD, approximately in order of their severity, reveals few if any unambiguous chronic effects. The industrial accident at Seveso in 1976 exposed some 37,000 people of all ages to considerable amounts of TCDD. A relatively small number of them showed transient effects such as chloracne (184 cases, 164 of them children under the age of 15), headaches and digestive upsets, but no long-term effects such as birth defects and chromosomal damage have been identified. It is too early to tell whether the incidence of cancer is abnormal.

An accident in a Monsanto plant in Nitro, W.Va., in 1949 exposed more than 200 workers to TCDD. Of 122 who developed chloracne, 121 were monitored for the next 30 years. The total number of deaths in that group did not differ significantly from that expected in the population at large, and there were no excess deaths due to cancer or diseases of the circulatory system. Similar findings have been made after other industrial accidents, except for two in which an excess of deaths from cancer was found in small groups of the people exposed.

A particular type of cancer (soft-tissue sarcoma, a generic term for more than 100 different types of rare cancer) has become a focus of concern because of a survey of Swedish forestry workers by Lennart Hardell of the University of Umea. He concluded that their exposure to 2,4,5-T (and thus to TCDD) had caused six times the normal incidence of soft-tissue sarcoma.

This study led to an investigation of chemical-plant workers in the U.S. who had been exposed to 2,4,5-T and other chemicals. Seven apparent cases of soft-tissue sarcoma were discovered, raising the level of concern substantially. Subsequent events have emphasized the difficulties in accurate diagnosis of soft-tissue sarcoma and in accurate identification of exposed individuals. At a conference in 1983 Marilyn A. Fingerhut of the National Institute of Occupational Safety and Health reported that two of the seven people had in fact died of cancers other than soft-tissue sarcoma. Moreover, the exposure of three others to TCDD could not be documented. Such findings fall far short of being hard evidence for the proposition that TCDD causes soft-tissue sarcoma.

Other studies also fail to support Hardell's hypothesis. In the state of Washington no consistent pattern of death due to soft-tissue sarcoma was found among occupations in which workers would have been exposed to TCDD. A study in Finland found no cases of the disease among 1,900 people who applied herbicides, nor was their death rate from any natural cause different from that of the total male population in Finland. The U.S. Air Force, in its Ranch Hand study of about 1,200 military personnel who sprayed Agent Orange in Vietnam, found no cases of soft-tissue sarcoma. Finally, examinations by the Veterans Administration of 85,000 self-selected veterans showed fewer cases of these cancers than the national average would suggest.

Reproductive effects are also a subject of concern because of the animal findings. The most celebrated case alleging such effects in humans is commonly known as the Alsea II study, made by the U.S. Environmental Protection Agency. The study reported a link between the spraying of 2,4,5-T on foliage and spontaneous abortion among pregnant women in Alsea, Ore.

This study has come in for much criticism, notably by an interdisciplinary group at Oregon State University. The group concluded that an association between herbicide spraying and spontaneous abortion could not be shown from the data relied on by the agency. Other studies—in Australia, Hungary, New Zealand and the U.S.—failed to find a link between the use of 2,4,5-T and birth defects.

Because of the extreme acute toxicity and the multiple chronic effects of TCDD in animals, regulatory agencies have had to consider what to do in order to protect people from exposure to the material. Such agencies must extrapolate animal data to human beings in all but a few instances, in spite of the fact that the validity of this type of extrapolation has not been ascertained. Compounding the difficulty is the lack of a simple, accurate method for determining whether and at what level TCDD occurs in the tissues of exposed individuals. (The present test requires a surgical procedure to obtain samples of the liver and fat tissues where TCDD resides.) Without such information a dose-response relation cannot be established.

TCDD has been called the most toxic synthetic chemical known to man. If its acute toxicity to the guinea pig, and even the rat and the mouse, is the criterion, the statement is probably correct. If its considerably lower toxicity to the hamster is the criterion, however, the statement would surely not be true. Yet there is no need to quibble: TCDD is unquestionably a chemical of supreme toxicity to experimental animals. Moreover, severe chronic effects from low dosages have also been demonstrated in experimental animals. Therefore the concern about its effects on human health and the environment is understandable.

When toxic chemicals are at issue, a regulatory agency has few options beyond extrapolating animal data to humans. Yet health effects on humans are rarely proved in the case of environmental chemicals to which the public is variably exposed at subacute levels that can only be estimated (and then only in the crudest approximation). A case in point is aflatoxin, the product of a mold that develops commonly in stored oilseed crops such as peanuts. In animal tests aflatoxin is one of the most potent carcinogens known, but it has not yet been proved to have this effect in human beings.

Diversity in reaction to stimuli is a hallmark of biological organisms. Reactions to toxins are no exception to the general rule. People may be more, less or equally sensitive to a given toxin than an experimental animal is. Extrapolation is neither art nor science; it is simply the most rational way to assess a hazard in the absence of definitive data. Hence regulatory actions continue to be based on the animal data even when the human data, although they are not definitive, may be sufficiently compelling to allow a scientific judgment that the hazard to people has been overestimated.

That appears to be the case with TCDD. Investigators are in general agreement that TCDD is less toxic to humans than it is to experimental animals, but the available information is not sufficiently compelling to stimulate a change in regulatory posture toward either more or less restriction of exposure to the material. I suspect that the direct evidence of TCDD's effects on humans will never be either more or less compelling than it now is.

The public's perception of a toxin is an important determinant of the posture taken by a regulatory agency. The public has heard a great deal about both the acute and the chronic effects of TCDD on experimental animals but little about the substantial body of data showing that human beings are less sensitive. The initial reports of TCDD's acute toxicity, followed by reports of its carcogenicity and reproductive effects, have instilled a public fear that probably cannot be dispelled even by adequate information about the countervailing experience with human beings. The regulatory agency is therefore left in the position of having to deal with not only the available evidence but also the public's fear.

The U.S. Environmental Protection Agency has responded to the public's fear with a number of regulations intended to control the formation and release of TCDD and to limit individual exposure to it. Those regulations could be made stronger or weaker on the basis of new evidence. What the agency has not done—and might be said to have a responsibility to do—is to try to dispel the public's fear on the basis of the evidence that exposure to low concentrations of TCDD in the environment appears not to have serious chronic effects on human beings.

The TCDD case is further exacerbated by its relation to the defoliation program in Vietnam, an unpopular program in an unpopular war. The many and diverse health effects alleged by Vietnam veterans to have been caused by exposure to Agent Orange have been widely publicized. The public is generally aware that the complaints were settled out of court for $180 million, and many people believe the settlement was an admission of guilt by the chemical companies that manufactured Agent Orange. Apparently few people know of Federal Judge Jack B. Weinstein's statement to the attorneys for the plaintiffs that "in no case have you shown causality for the health effects alleged."

A troublesome matter exemplified by the TCDD issue is the appropriate utilization of scientific resources. A. L. Young of the Office of Science and Technology Policy has calculated that more than a billion dollars will have been spent by the Federal Government for research and other dioxin-related matters before all the major studies now in progress have been completed. Additional expenditures of both time and money have been made by chemical companies, private organizations and government agencies. The total outlay is a tremendous amount for an issue of questionable importance.

Two years ago a conference on dioxin at Michigan State concluded that the TCDD case is relatively less important than a number of other issues and that the nation's limited scientific resources should be devoted to the issues posing a greater threat. On the basis of the evidence turned up so far, the conclusion is still valid.

IV. NUCLEAR WASTE

EDITOR'S INTRODUCTION

Perhaps no aspect of waste disposal is more alarming than that of nuclear waste. The nuclear power industry has been in existence for four decades, and despite alarm caused by a partial meltdown at the Three Mile Island facility in Pennsylvania, no significant releases of radioactivity from nuclear power plants have taken place in the U.S. Nevertheless, the wisdom of expanding nuclear power is still debated, and the problem of disposing of nuclear waste in particular has plagued the industry. No satisfactory solution to permanent nuclear waste storage has yet been found. As more nuclear plants come on line and more waste accumulates, the problem will continue to grow. This ever-growing pile of almost inconceivably deadly material, with a life span measured in hundreds of thousands of years, is a threat to the future of the planet.

In the first article in this section, "Nuclear Energy Facts: Questions and Answers," a booklet prepared by the American Nuclear Society, reassurances are provided on every phase of the operation of nuclear power plants, as well as the industry's present and future plans for waste disposal, including the selection of a permanent storage repository in the not-too-distant future. A following article, "Bad News from Britain," by the novelist Marilynne Robinson, focuses upon mismanagement of nuclear waste in England but more generally raises questions about the credibility of the scientific community.

In a third article in this section, Tom Yulsman, writing in *Science Digest,* addresses the question of a permanent disposal site for our ever-growing volume of nuclear waste. As Yulsman explains, high-level waste has been stored temporarily at nuclear plants themselves, awaiting deposit in a permanent shelter planned for 1998. Yet delay has followed delay as feasibility studies have produced no certainty about the safety of this material over the course of centuries. Yulsman points out that a subterranean depository requires an exact knowledge of geological stresses and change that cannot be predicted, since previous patterns of change may not repeat themselves. In the meantime, the nation is running out of temporary storage space, and some action will have to be taken soon.

Two following articles examine the problem of low-level nuclear waste, consisting of apparel, tools, etc. that have been exposed to radioactivity and are stored in plant annexes and then in regional depositories. In the first of these articles, written by Gale Warner for *Sierra*, the background of disposal (in the 1950s, put in drums and dumped overboard from navy ships; in the 60s and 70s, deposited in landfills) is brought out, as well as the incidence of leakage from low-level waste sites (at Maxey Flats, Kentucky, plutonium was detected more than a mile away). During the last nine years, low-level waste sites have been concentrated in only a few areas, but congressional legislation enacted in 1980 (the Low-Level Radioactive Waste Policy Act) requires the separate states to provide for their own waste by 1993, a deadline that will involve crucial judgments affecting future environmental safety. In the related article, in *Science Digest*, Susan Q. Stranahan concludes that this legislation was hasty and ill-advised. Given too little time to plan wisely, the states may well stress appearances and speed rather than technological excellence—continue to mix high-level with low-level waste, and select shallow landfill disposal over safer (but more costly) above-ground depositories. Finally, Cynthia Pollock, in a revealing article in *Environment*, reports on the decommissioning of nuclear power plants, which have a life expectancy, because of radioactive buildup, of only 30–40 years. Their dismantling or "entombment" will be vastly more expensive than originally estimated, as much as $3 billion or more for each plant, plus the cost, because of a terrorist threat, of guarding the sealed plants indefinitely. Taken together these articles raise the lingering question if the expansion of the nuclear energy industry is really the most prudent course that we should be following.

NUCLEAR ENERGY FACTS:
QUESTIONS AND ANSWERS[1]

Do we really need nuclear power?
Yes. A look at our fossil fuel resources and our increasing dependence on foreign supplies will explain why. Oil and natural gas supplies are running out and becoming increasingly costly. Because of the long time it takes to build a new plant (about 12–14 years) and the even longer time it takes to develop new technologies, we have to start making some hard decisions right now. Unlike oil and gas, our coal supply is relatively abundant, and also unlike oil and gas, the generation of electricity is its major use. With world energy use growing as the populations expand and economies develop only coal and nuclear energy are now available in large supply. Coal has its own problems in terms of air pollution. However, coal will be needed in the future as a substitute for oil and natural gas. There are few uses for the uranium other than for the production of nuclear energy; the quantities necessary are much smaller; and nuclear power plants are cleaner. Thus, nuclear energy is needed to take over an increasing share of electrical production if we are to become less dependent on foreign oil.

Won't conservation of fossil fuel resources and generated electricity make nuclear power unnecessary? Why do we need nuclear?
Conservation is important and must be encouraged. However, conservation alone will not solve the energy problem. We are unlikely to want zero economic growth, if for no other reason because our population growth will not achieve equilibrium until the 21st century, even with continuation of low birth rates. Even moderate population and economic growth will require increased production of energy.

Nuclear is important because the recoverable energy released from the nuclear reaction for the generation of electricity is about 68,400 times greater than the energy recoverable from

[1]Reprint of American Nuclear Society booklet. Reprinted by permission from American Nuclear Society. Copyright © 1985 by American Nuclear Society.

burning an equivalent weight of coal in a fossil fuel plant. Ton for ton, uranium ore yields over 50 times the energy of coal. The amount of uranium required each year for a nuclear plant is extremely small (about 30 tons) compared with the amounts of fossil fuels required for the same size plant (2.3 million tons of coal, or 10 million barrels of oil, or 64 billion cubic feet of gas). In addition, 97% of the 30 tons of uranium fuel is reusable. It's cheaper than any way you can think of.

How about the costs of power? Isn't nuclear power more expensive than the other sources?
Costs for building nuclear plants have greatly increased due to regulations and court delays: many of these delays have done nothing to improve reactor safety or reliability. Coal-fired plants are also facing increasing costs due to regulations, so nuclear plants will remain cost effective. Over the long term, nuclear generation of electricity can produce large savings over coal generation because of the difference in fuel costs and transportation expense.

Why should we depend on nuclear power when some critics say the reactors are unreliable and uneconomic?
The reliability of nuclear plants, on the average, is as good as, or better than, coal plants of the same age and size—within 1 or 2 percentage points of each other. This is true both for number of hours the plants are available each month and for the percentage of generating capacity actually achieved. A few nuclear plants, especially some older ones, have had operating problems that delayed their startups or reduced their operating times. Generally, the problems (in both coal-burning and nuclear plants) are with conventional steam equipment and are unrelated to the nuclear reactor.

A lot of people are worried about radiation from nuclear power plants. How much radiation do I get from the generation of nuclear powered electricity?
Very little. As a gauge, a person in the U.S. receives on an average 180 millirem (mrem) per year from all sources. A millirem is a measurement of the effect of radiation on living tissue. Most is natural: from soil, water, rocks, building materials, food. Since 1970 radiation from all commercial nuclear energy averaged

0.01 millirem for each person in the U.S. In the year 2000, assuming nuclear energy becomes a dominant source of electricity, the average citizen will receive an estimated yearly dose of less than 1 mrem from commercial nuclear energy. Those living near nuclear power plants will receive less than 5 mrem per year.

If small amounts of long-lived radioactive materials are released to the environment, isn't there a buildup to dangerously significant levels over a long period of time?
No. Government agencies limit releases from nuclear power plants and the entire fuel cycle. As for the risk or danger, the levels of releases from nuclear plants may shorten the human life as much as 24 seconds. You can place this in perspective by realizing that being 25% overweight decreases the lifespan by 3.6 years, smoking a pack of cigarettes a day can decrease life by 7 years, and living in the city rather than in a rural area can decrease it by 5 years.

Why is any release of radioactivity permitted?
It is as impossible to have zero releases from nuclear plants as it is to have zero releases of pollutants from any industrial process. What is done is to assure that any releases are well below the levels of significant environmental or human health effects; these limits are set by national and international groups and are based on vast quantities of data collected for over 50 years. This attention to releases has been observed in the nuclear power industry from its inception. In contrast, most other technologies were fully developed and in use before pollution control was required or achieved. Radioactive materials are routinely released, with no controls, from coal-burning power plants; this radioactivity comes from minerals that are a natural part of the coal.

Has any person in the United States ever been exposed to an overdose of radiation from commercial nuclear power plants?
The public has never been exposed to radiation levels above the annual dose limits established by international standards. Rare cases of researchers and workers being overexposed have occurred, mostly in early days of nuclear science. Workers in commercial nuclear power production are, in fact, protected with extensive precautions to prevent exposure which might adversely affect them even many years later.

What are power plant operators doing about thermal pollution?
Power plant operators must comply with strict federal, state, and local regulations limiting the amount of unused heat (in the form of water, liquid, or vapor) power plants can release into the environment. Most states limit the rise in water temperature at the plant's heated-water discharge to less than 5°F. The Environmental Protection Agency requires either a system using ponds or towers for precooling the water or proof that direct discharges of power plant heat will not damage the ecological balance of the body of water or the plant and animal life dependent on it. Power plant operators must monitor the discharges and bodies of water to demonstrate continually that the cooling systems are functioning properly with no resulting damage to the environment.

How would an earthquake affect a nuclear power plant located near the quake center?
Nuclear power plants are generally located away from earthquake-prone areas and are carefully designed to withstand an earthquake if one should occur. This applies to all parts of the United States and in other countries as well. Each plant must be able to withstand the maximum earthquake motion that could be expected at its site and be able to shut down safely. Unlike fossil plants, which are not designed to the same exacting standards for earthquake resistance, nuclear plants can be expected to continue operating during a moderate earthquake. Nuclear plants near earthquake epicenters in California and Japan have successfully withstood earthquakes of magnitudes as great as 6.5 on the Richter scale.

What kinds of accidents can occur with nuclear power plants?
First of all, nuclear power plants cannot explode like an atomic bomb because they do not contain the necessary concentration of fissionable material. Reactors are designed and constructed to withstand the kinds of accidents that could occur. Most often mentioned is an accident that could cause loss of the coolant water as the result of a rupture of a large pipe. This has **never** happened in a commercial reactor. Despite the damage to the core and the small releases of radiation at TMI, the defense-in-depth design philosophy prevented any serious damage to the rest of the plant or to the health and safety of the public.

How is the public protected against various potential hazards?
In a nuclear power plant, the potential hazard is the large amount of radioactive products created in the fuel during the fission process that generates the heat to make the electricity. Plants are designed to contain these fission products in the event of an accident. This is done with multiple physical barriers, beginning with the fuel itself and proceeding to metal and concrete barriers around the reactor core and the entire reactor system..Each plant is also surrounded by an area from which the public is excluded.

Many safety systems, including backup systems, are designed to shut down the reactor if the plant begins to operate abnormally. All people who work in the reactor's control room go through rigorous training and must be licensed by the U.S. Nuclear Regulatory Commission (NRC). Operators must follow written instructions precisely to assure safe and dependable nuclear plant performance.

Following TMI, industry established the Institute of Nuclear Power Operations (INPO) to improve further the standards for reactor operator training.

Doesn't the 1979 accident at Three Mile Island prove that nuclear power is too dangerous?
No. The accident at Three Mile Island (TMI) had a traumatic effect on some of the people in the vicinity of the plant because of a fear of a meltdown of the nuclear fuel. Several expert studies, including the Kemeny Report commissioned by President Carter, agree that a meltdown would not happen and that the defense-in-depth safety design worked to protect the public health and safety. They also agree that the effects on the population in the vicinity of Three Mile Island from radioactive releases during the accident, if any, will certainly be nonmeasurable and nondetectable. The accident was serious, but no lives were lost, no one was physically harmed or is likely to suffer future ill effects. The industry and government are diligently searching for and applying "lessons learned" from the accident at TMI to plant and reactor design, operator training, communication, and regulation.

How good is the nuclear power plant safety record?
In over 25 years of nuclear reactor operation by electric utilities in the U.S., no property damage or injury to the public or operating personnel has ever been caused by radiation from these facili-

ties. U.S. naval vessels have a similar accident-free record for an even larger number of nuclear plants. Studies show that public risks of adverse health effects from nuclear plants are less than from power plants using other kinds of fuel. This remains true even after the accident at Three Mile Island, Pennsylvania, in 1979. The greatest injury resulting from that accident was mental anguish caused by fear.

How are nuclear power plants licensed and regulated?
Before any plant can go into service, the utility must obtain many different licenses and operating permits from federal, state, and local agencies. First, a construction permit from the U.S. Nuclear Regulatory Commission is required. Then, after construction is completed, an operating license must be obtained. After the plant goes into service, the NRC carries out on-site inspections to assure that the plant is operated according to its license. Each utility monitors its plants for radioactive discharges and the records are audited by the NRC and the Environmental Protection Agency. Abnormal operations or conditions are reported to these agencies. Independent inspections are also made by insurance companies (see the following question and answer).

Why is there a "nuclear exclusion" clause in homeowner's property loss insurance?
Most homeowner's policies have clauses excluding coverage for nuclear damage as well as for various natural disasters. Nuclear exclusion clauses exist because such coverage is channeled into nuclear insurance pools. Groups of private insurance companies, together with Federal Idemnity and the nuclear utilities, supply the legally required liability coverage for all nuclear facilities that could cause damage to the public, to a maximum of $560 million. The nuclear utilities pay appropriate premiums for both the private insurance pool coverage and the Federal Indemnity. The private insurers have committed $160 million for damage to the public, plus $30 million of secondary liability coverage, and $300 million of property coverage, thus making a current commitment of $490 million for homeowner's property insurance from the insurance industry.

Isn't the Price-Anderson Act a subsidy to the nuclear industry? Without it, wouldn't the industry cease operation?

No. First, there is nothing new about Federal Indemnity programs. They already include crop insurance, national flood insurance, medicare, and other large-scale programs. Second, the nuclear utilities have and are paying premiums to the Federal Government for the indemnity (the government has never had to pay a claim under the coverage). Since the indemnity portion of the Price-Anderson Act is being phased out as a responsibility of the government and is being assumed by the nuclear utilities, it is apparent that the loss of this indemnity is not causing the industry to cease operations. Three Mile Island demonstrated the need to cover the cost of power purchased to replace the power lost as a result of the accident; industry has now established Nuclear Energy Insurance Limited (NEIL) to help pay for replacement power in future accidents that result in loss of power production.

Why hasn't the waste disposal problem been resolved?

Several plans for handling these wastes *have* been worked out. Scientists, through many years of research, have developed alternate ways to contain and store radioactive waste safely. The need for permanent waste disposal exists even if commercial nuclear power is not continued. Wastes from defense programs at Hanford, Washington, Savannah River, South Carolina, and Idaho National Engineering Laboratory, Idaho, already far exceed the quantities that will be produced by the year 2000 from commercial nuclear power plants.

The volume of wastes is readily manageable. If the liquid wastes from fuel reprocessing are "cooled," converted to stable solid form, and permanently stored at a federal repository, all the nuclear waste—including the low-level waste—from the entire U.S. nuclear power industry until the year 2000 would fit into a cube 250 feet on a side. The high-level portion of the radioactive wastes would take up a cube 50 feet on a side within the 250-foot block.

The National Academy of Sciences and other noted scientific organizations have stated clearly that the technology exists now for safe disposal of radioactive waste. The problem, then, is a political one, centering around federal licensing of facilities and states' rights in siting. A demonstration of the feasibility of safe waste storage requires action by Congress to mandate immediate construction of such facilities.

Is it right for us to leave a "legacy" of radioactive wastes as a hazard for future generations?

The consequences of alternatives to nuclear power should be pondered in determining if we consider it morally acceptable to establish carefully guarded and monitored repositories for the high-level radioactive wastes. We may burn up all the fossil fuels that *ever existed on earth* in only a few hundred years just to sustain industry, to feed current world populations, and to satisfy the growing demand around the world for an adequate standard of living with hope of improvement. Should we leave our descendants without fossil materials (coal, oil, gas) from which to extract fertilizers, medicines, and plastics because we elected to burn these fossil materials instead of using nuclear power? Is it fair to leave them without adequate energy to provide employment and the chance to choose their own lifestyles? In a democratic society, a public consensus based on informed opinion must answer these questions.

What happens to low-level wastes?

At present, all low-level radioactive wastes are packaged and shipped for land burial to government licensed low-level waste storage sites. Transportation and burial of the wastes are regulated and controlled by two primary federal agencies, the Department of Transportation (DOT) and the Nuclear Regulatory Commission (NRC).

Low-level radioactive wastes from a nuclear plant are composed of resins from water purification filters, lab supplies, and trash. The typical nuclear power plant generates about 30,000 cubic feet per year of low-level wastes, equal to the contents of 4,000 55-gallon drums. (The high-level wastes are composed of the radioactive fission products contained in the used fuel from the nuclear reactor, about 3% of the total fuel.)

A large portion of the country's low-level radioactive wastes comes from hospitals in the form of empty containers from which the radioactive material has been drawn and put into human beings for diagnostic purposes or treatment. Nuclear medicine in the form of radioactive isotopes is used thousands of times daily in U.S. hospitals. The medical low-level wastes are also sent to government licensed low-level waste storage sites for burial.

How difficult would it be to build a nuclear bomb if you could obtain the basic materials?

You'd also need the knowledge, hardware, special materials, and specific abilities. You *cannot make* an explosive device with the enriched uranium that is used as fuel for power reactors. Uranium enrichment is a difficult and costly process that could be carried out only by a nation with large resources and trained people. Plutonium in the spent nuclear fuel removed from reactors must first be separated from other materials. (From a practical point of view, theft of the nuclear fuel is difficult. For example, a crude bomb using 25 pounds of plutonium would require the theft of about 1,500 pounds of radioactive fuel, removed from 12-foot-long reactor fuel elements transported in a shipping container weighing up to 100 tons.) Then, before a nuclear bomb with the plutonium could explode, precise amounts would have to be brought together with just as precise shaping, sequencing, and timing.

Can't the disperal of radioactive materials be an effective terrorist device?

Perhaps, in theory. Some scenarios say the most likely radioactive material used as a threat would be plutonium. Although plutonium is radiologically toxic when small quantities are inhaled and ingested, its effects do not occur until 25 to 45 years after exposure. In preference to materials that have delayed effects, like plutonium, terrorists could be expected to choose the psychological impact of immediate damage, such as that caused by bombs and fires. Also, terrorists would have great difficulty obtaining plutonium, while equally or more toxic chemical and biological poisons are much more readily available. The International Fuel Cycle Evaluation made in 1978–80 determined that strict control and safeguards are the best way to prevent unauthorized access to nuclear materials.

Doesn't selling nuclear power plant components abroad, where we can't control their use, increase the kind of danger we've been talking about?

Nuclear power plants are being built and sold around the world by France, Germany, Japan, and other nations. The U.S. can no longer control the use of this valuable resource. Government regulations and restrictions of 1978–79 on exports literally removed the U.S. from the world nuclear leadership; this further reduced the U.S. ability to control the spread of nuclear facilities. Since facilities for fuel reprocessing and uranium enrichment present the greatest opportunity for proliferation, some people believe that the U.S. should provide these services for other countries to help deter the transfer of nuclear materials into unstable areas of the world.

How toxic is plutonium? How many people have ingested plutonium and, as a result, died or developed cancer?
Plutonium, although recognized as hazardous, is *not* the "most toxic material known to man," as has been charged. Various organisms, chemicals and biological agents, such as snake venom, are lethal in smaller amounts and work faster than plutonium. However, while plutonium can be hazardous if inhaled or ingested, there are no known deaths attributable to plutonium poisoning. Of 25 persons who worked with plutonium at the Los Alamos Scientific Laboratory during World War II, not one has developed cancer since their exposure.

The nuclear industry says we need breeder reactors. Why?
Only an estimated 30 or 40 years' worth of uranium remains that can be readily and economically extracted from the ground for use in nuclear reactors. If we are not to use up this resource—the way we are using up our fossil fuel resources—we need to make more efficient use of our nuclear fuels. Worldwide, there is a strong agreement on developing the "breeder" reactor. Breeder reactors can turn the vast majority of uranium atoms (which are not fissionable) into fissionable fuel that can be recycled to produce more energy. Because they produce more fuel than they burn, such reactors are said to breed fuel. The breeder could increase the usable amount of uranium by more than 70 times, stretching 40 years' worth of uranium into many centuries' worth of fuel. In the U.S.A., non-fissionable uranium sitting in drums at government sites is equivalent to all of the estimated oil resources of the entire world if we utilize it in breeder reactors.

This source alone, worth more than $60 trillion dollars at today's oil prices, could provide our total electrical energy needs for several centuries.

Isn't it especially hazardous to ship nuclear fuel and high level wastes from one nuclear facility to another?
It is no more hazardous to ship nuclear fuel and high-level radioactive wastes than to ship many other materials we routinely transport all over the country, such as chlorine. Three types of hazards are considered in the transportation of nuclear fuel and wastes—radioactivity, ability to form a critical mass, and theft. The first is controlled by preventing the release of highly radioactive materials, even in the most severe accident. To ensure this, shipping casks have been deliberately subjected to collisions between trucks and trains, and trucks and a concrete wall. In all cases, films document that there was not enough damage to the shipping cask to permit release of any radioactive contents. The second is controlled by the use of appropriate shielding and by separating materials. Theft is preventable by a combination of impenetrable shipping materials and security personnel. All shipments are subject to strict federal regulation.

How do the risks from nuclear power compare with other everyday risks?
Nuclear power offers less risk. This sort of comparison was examined in the many safety studies, most notably the Rasmussen Report. Events such as air crashes and explosions have more than 100,000 times the chance of killing 10 people than 100 nuclear plants, the study concluded. It also found that a dam failure has 10,000 times the chance of killing 1,000 people than 100 nuclear plants. Far greater consequences are calculated for natural disasters—earthquakes are 2,000 times more likely to kill 10 people than 100 nuclear plants. Hurricanes are 60,000 times as likely to kill 1,000 people as 100 nuclear plants.

Can't we use solar and fusion power instead of nuclear power?
Solar, fusion, and other power sources—including hydroelectric, wind, and geothermal power—are limited in future expansion by location and natural conditions or the need for a long period for development of the technology. Hydroelectric power in the U.S. is just about fully exploited. Solar power is not expected to be adaptable to large, central station electric power production in this century, if ever. However, any contribution from solar energy would be helpful. Utilities' demands peak during the day even when the sun shines. Additional oil or coal must be burned to assist base-load nuclear power plants.

The fusion reaction, so far created only in the laboratory, has a long way to go before it can be demonstrated on a large or economic scale. Even in certain areas where windpower might be harnessed, relatively small amounts of electricity will be produced. More geothermal power could be developed in a few specific areas of the West where sources occur, but only as a minor contribution to the nation's total electrical needs. In addition, the safety and environmental effects of those alternative energies have not yet been fully assessed.

Aren't we really talking about the quality of life? Who decides if the benefits are worth the risks?
Ideally, in our democratic society the decisions are made by consensus or by majority decision. One aspect of this can be seen by the voting in 1976 in various states on the issue of "safe power," where majorities of the voters decided additional restrictions on nuclear power plants were not desirable. The public in general has not turned against nuclear power since the accident at Three Mile Island. Voters in Maine in 1980 refused to shut down Maine's only operating nuclear plant.

More subtle, societal consensus depends on a "trade-off" involving not only health and safety, but also esthetics, quality of life, and ecological balance. Sometimes societal consensus is embodied in law, such as the National Environmental Policy Act, or in court decisions. Or, it can be built into a licensing process, as is the case with nuclear power plant construction and operation, where public input is a vital part of the process.

What is our responsibility?
Some affluent people contend that there is too much technology today; that it is the basis for many of society's ills, and that we must, therefore, put a stop to further development. Exaggerated are the problems caused by technology while forgotten are the truly large societal problems that technology has helped solve. For example, as we condemn Detroit and auto emissions, let us not forget New York early in this century with 150,000 horses in the street and their emissions.

Throughout history, mankind has had two basic choices when confronted with a problem that technology could solve. The first choice has been to ignore the promise of technology and endure the problem—a choice which has invariably led to reduce comfort, well-being, and security. The second choice has been to put technology to work to solve the problem—a choice which has assured increasing prosperity, opportunity, and hope for all mankind.

It is essential that any decision be based on scientific facts and not on dreams. It is the responsibility of the scientific community with expertise on energy technologies to inform the public on the facts. It is your responsibility to become informed and to choose on that basis.

BAD NEWS FROM BRITAIN[2]

On the coast of Cumbria, in the English Lake District, there is a nuclear reprocessing plant called Sellafield, formerly Windscale, that daily pumps up to a million gallons of radioactive waste down a mile and a half of pipeline, into the Irish Sea. It has done this for thirty-five years. The waste contains cesium and ruthenium and strontium, and uranium, and plutonium. Estimates published in the London *Times* and in the Sunday *Observer* are that a quarter of a ton of plutonium has passed into the sea through this pipeline—enough, in theory, according to the *Times*, to kill 250 million people; much more than enough, in theory, according to the *Observer*, to destroy the population of the world. The plant was designed on the assumption that radioactive wastes would lie harmlessly on the sea floor. That assumption proved false, but the plant has continued to operate in the hope that radioactive contamination may not be so very harmful, after all. If this hope is misguided, too, then Britain, in a time of peace, has silently, needlessly, passionlessly, visited upon us all a calamity equal to the worst we fear.

Everything factual that I will relate in this article I learned from reading the British press or watching British television. But

[2]Reprint of an article by Marilynne Robinson, novelist. Reprinted by permission of the author. Copyright © 1985. This article was first published in *Harper's*, 270:65-72. F. '85.

it would not be accurate to say that I know, more or less, what a reasonably informed Briton knows about these things, because there is a passivity and credulousness in informed British opinion that neutralizes the power of facts to astonish.

To understand what I will tell you, you must imagine a country where, though the carcinogenic properties of radioactivity in general and of plutonium in particular are gravely conceded, it is considered reasonable, in the best sense, to permit the release of both of these into the environment until the precise nature of their effect is understood. This notion of reasonableness is, I think, extremely local, but the consequences of such thinking are felt in many places. The Danes object to plutonium on their beaches, as do the Dutch. And of course the Irish, a volatile people at the best of times, are now very much exercised by elevated rates of childhood cancer and Down's syndrome along their eastern coast. They have leaped to the very conclusion the British find too hasty—that the contamination of the environment by known carcinogens is detrimental to the public health. No one disputes that the contamination of these coasts is surely and exclusively owing to British reasonableness, since the Irish have not developed nuclear energy—nor have the Danes, who consider it unsafe—and since the only other fuel reprocessing plant known to release wastes into the sea, at Cap de la Hague in France, releases only one percent of the radioactivity that enters the sea from Sellafield.

When I realized what I was reading, I began to clip out articles every day and save them, and I have brought them home, knowing that my uncorroborated word could not be credited. Travelers to unknown regions must bring back proof of the marvels they have seen. Perhaps the most incredible part of this story is that it has fallen to me to tell it. American scholars and scientists go to Britain in platoons. Many live there. Probably all of them look at the *Guardian* now and then, or the *Times*. Perhaps most of them are more competent to understand what they read there than I am, better schooled in such matters as the particular virulence of plutonium, or the special fragility of the sea. No one had ever hinted to me that for thirty-five years Britain has knowingly befouled itself and its neighbors with radiation, and nothing I had heard or read had prepared me to discover a historical and political context for which the one vivid instance of Sellafield could well serve as an emblem. Yet Sellafield does not depart from, but in fact epitomizes, British environmental practice. This is only to say, read on. This is a tale of wonders.

In November of 1983 a family was walking along the beach near Sellafield—it is a major tourist and recreational area—when a scientist who worked at the plant stopped to tell them that they should not let their children play there. They were shocked, of course, and raised questions, and sent a letter to their MP. The scientist was fired, amid official mutterings about his having committed an impropriety in disclosing this information. No doubt he had violated the Official Secrets Act, though so far as I know the matter was not couched in those terms. British workers in significant nationalized industries—for example, British Aerospace, the postal system, and the nuclear industry—are obliged to sign the act, which imposes on them fines and imprisonment if they reveal without authorization information acquired in the course of their work. Only death can release them from this contract. Employees of private industries are in the same position, for all intents and purposes, since the unauthorized use of privately held information is prosecuted as theft. In the democratic kingdom, the exercise of judgment and conscience is the exclusive prerogative of the great.

But I digress. Though the renegade employee was dismissed, the issue of the safety of the beaches was called to public attention, with a series of consequences. A woman who lived in a village near Sellafield sent a bag of dust from her vacuum cleaner to a professor in Pittsburgh, who found that it contained plutonium. Divers from British Greenpeace tried to close the pipeline but were unable to do so because the shape of its mouth had been altered. They discovered an oily scum on the water that sent the needles of their Geiger counters off the scales. The divers and their boat had to be decontaminated. The radioactive slick was said to be the consequence of an error at the plant that had disgorged a radioactive solvent into the sea—an accident that, unlike the normal functioning of the plant, raised questions of competence and culpability. That is to say, this matter was put into the hands of the Director of Public Prosecutions, and quite appropriately. However, it is a curious feature of British law and practice that silence descends around any issue that is about to become the subject of legal action. A judge may remove this restriction in particular cases; murder trials, for example, are reported in lascivious detail. But a newspaper that publishes anything

relating to matters prohibited as sub judice is subject to cata-
strophic fines. The manufacturers of thalidomide, the sedative
that caused many British children to be born without limbs, kept
the question of their liability before the courts for seventeen
years, and therefore unresolved and out of public awareness, until the *Sunday Times* defied the law and broke the story. The newspaper took its case to the European Court of Human Rights, and
won, but this has had no effect on British law or practice. British
justice, which is cousin to British reasonableness, grows squea-
mish at the thought that the legal process should be adulterated
by publicity.

As a third consequence of the attention drawn to Sellafield,
Yorkshire Television sent a team there to look into worker safety.
The team discovered that children in the villages surrounding
the plant suffered leukemia at a rate ten times the national average. This revelation fueled public anxiety to such an extent that
the government was obliged to appoint a commission to investigate. It recently published its conclusions in the so-called Black
report, named for Sir Douglas Black, president of the British
Medical Association and the commission's head and spokesman.
Dr. Black startled some by assuring a television interviewer that
people fear radioactivity now just as they feared electricity one
hundred years ago.

The report offers "a qualified reassurance" to those concerned about a possible health hazard in the area. The *Guardian*
said: "Recognising that radiation is the only established environmental cause of leukemia in children, 'within the limits of present
knowledge,' the Black team calls for new studies to provide additional potential insights." Again according to the *Guardian*,
"Despite the high rates of cancer close to Sellafield, the report
stresses: 'An observed association between two factors does not
prove a causal relationship.'" This is certainly true. And this is
the darling verity of the British government. Souls less doughty
than these might feel that exposure to radiation around Sellafield, together with an elevated cancer rate, testifies to a causal
relationship between these two factors, but we're not dealing with
a bunch of patsies here. In environmental issues, a standard of
proof is demanded that makes the Flat Earth Society look easy.

What do we have here? The better college sophomore has
learned that this world does not yield what we call "proof" of
anything. That so weighty an edifice as public policy should be

reared upon an epistemological abyss is truly among the world's marvels. Are these decision makers, known to wags as the Good and the Great, cynical connivers, imposing upon what can only be a frighteningly naive and credulous public? Or are they themselves also frighteningly naive? I cannot think of a third possibility. Whatever the cause of their behavior, its effect, like the effect of the Official Secrets Act and the contempt laws, is to shield government and public and private industries from suspicions of error or wrongdoing, and to blur, fudge, and frustrate questions of responsibility and liability.

You will note that the laws and practices and attitudes I describe here have existed over decades, and have persisted while governments rose and fell. For example, in 1974 the government passed the Control of Pollution Act. To have a proper understanding of "pollution" in this context it is essential to realize that in Britain, no legal control is exerted over agricultural chemicals or sprays. DDT is still in general use, as are aldrin and lindane. I know of no reason to imagine that policies toward industrial pollutants are any less indulgent in effect. Inspectors politely inform manufacturers of their intention to visit, so control of effluents can hardly be stringent. And we are not speaking here of soapsuds. In any case, part two of this Control of Pollution Act is now to be implemented, reports the *Guardian*. The article goes on to say, "The new measures are expected to have a big impact on the problems of Britain's dirty beaches." This seems to me a remarkably cheerful thought, considering that, to quote again, "the measures only apply to new sewage or trade effluent discharges, however. Existing discharges will continue, but 'consents' already granted will be subject to public scrutiny." Well, this looks to me like an act designed to confer legality on the very sources of pollution that already dirty Britain's beaches. However, the act must have a fang, if only a small one, because for ten years it was not implemented. Why? The article offers an explanation from William Waldegrave, Under-Secretary for the Environment, who "said that one of the factors that had held back successive governments was the fear of increasing costs to industry."

How is one to understand the degradation of the sea and earth and air of the British homeland by people who use the word *British* the way others of us use the words *good,* and *just,* and *proud,* and *precious,* and *lovely,* and *clement,* and *humane?* No matter that these associations reflect and reinforce the complacency that allows the spoliation to go unchecked; still, surely they bespeak self-love, which should be some small corrective. I think ignorance must be a great part of the explanation—though ignorance so obdurate could be preserved only through an act of will.

The issue of Sellafield is complicated by the great skill the government has shown in turning accidents to good account. You will remember that the Greenpeace divers surfaced through highly radioactive slime. If they had not had Geiger counters with them, no one would have known that an accident had taken place. Ergo, one cannot know that *other* accidents have *not* taken place. From which it follows that these accidents, and not the normal functioning of the plant, might be responsible for the cancers and other difficulties and embarrassments. As the *Guardian* said, in its sober and respectful paraphrase of this startling document, the Black report, "The possibility of unplanned and undetected discharges having delivered significant doses of radiations to humans via an unsuspected route could not be entirely excluded." The implication of all this is that the plant can be repaired, improved, and monitored, and then the hazards will go away. Number eight on the list of ten recommendations by the Black inquiry team suggests that "attention [be] paid to upper authorized limits of radioactive discharges over short periods of time; to removal of solvent from discharges and adequacy of filter systems"—in other words, if occasional splurges are avoided, the level of radioactivity will remain safe and constant. That might well be true, if the substances put in the sea decayed. But as the *Observer* has noted, plutonium remains toxic for at least 100,000 years.

Another accident that has had great effect on the way this affair has been managed is the fact that Yorkshire Television focused its attention on leukemia among local children. This is understandable, since the deaths of children are particularly vivid and painful to consider. But the limiting of the discussion to childhood cancer in the Black report is clearly arbitrary and possibly opportunistic. Seascale, the village nearest the plant, where seven children have died of leukemia in a period of ten years, has a population of 2,000. Children living there are said to have one chance in sixty of developing leukemia, but the sample is considered too small to be reliable—coincidence might account for the high incidence of the disease.

But why are we talking only about leukemia? I noted with interest, and added to my collection, a brief report about an inquest into the death of a Sellafield worker from bone cancer. An environmental group (not named) had pointed out that Dr. Geoffrey Schofield, the plant's chief medical officer, "did not mention the three most recent deaths from bone cancer at Sellafield." The article continues, "Dr. Schofield, quoting a 1981 report on mortality rates among British Nuclear Fuels workers at Sellafield, referred to four cases of myeloma, a bone cancer. These figures over the period 1948 to 1980 were comparable with national figures. Since that report three more workers have died from myeloma and a fourth appears to have contracted the disease." How do these cancer deaths relate to the cancer deaths among children in the area? Doesn't the concentration on the young actually focus attention on that portion of the population least likely to have developed cancer?

But officially preferred hypotheses are invoked to preclude lines of inquiry that might produce data that would discredit them. What harm could there be in checking for lung cancer deaths in areas downwind of Windscale?* These would certainly be equally relevant to the question of public safety, the real issue here.

The conclusion reached by James Cutler, the Yorkshire Television producer who first made public the high incidence of leukemia in Seascale, and the great fear of the chairman of British Nuclear Fuels Limited, who really is named Con Allday, is that

*A striking feature in all this is the seeming difficulty of obtaining and interpreting information. One would think that a country with a national health service would enjoy centralized and continuous monitoring of health data. One would expect it to encourage preventive practices on both the public and the individual levels, if only on grounds of economy. But the British government has actually suppressed reports on alcoholism and on the relation of cardiovascular disease to diet—the second of these was leaked to the *Lancet*; the first, though joked about in the press, has been dubbed an Official Secret, and its findings may not be published. The British government saves money in the most direct way: by refusing to spend it. In the European Community only Greece spends a smaller share of its wealth on health care. Yet the British are proud of their health system. Margaret Thatcher is fond of saying they get "good value for money," and one often sees statements to the effect that indicators of general health show the British system outperforming the big spenders. If this is true—if, with poverty and unemployment and all the problems that extend them; if, with rampant abuse of alcohol and heroin, a polluted environment, and immunization policies so casual that Britain still has rubella epidemics; if, with a slow rate of decline in cigarette smoking and rates of breast and lung cancer at or near the top of the charts—if Britain still does better than countries that devote more generous portions of larger resources to populations whose conditions of life are distinctly more consistent with well-being, then the National Health Service beggars any praise.

anxiety among the public signals a defeat for proponents of nuclear power. Now, I think nuclear power has proved to be a terrible idea, but I do not think the practices associated with Sellafield should ever be spoken of as if they were characteristic and inevitable aspects of its development. To do so would be to obscure the special questions of competence, of morality—of sanity, one might say—that Sellafield so vividly poses. But as I said earlier, I do not wish to imply that what has been done at Sellafield departs radically from the *British* nuclear establishment's behavior. Ninety percent of the nuclear material that has been dumped in the sea worldwide has been dumped by the British. They have deposited it off the coasts of Spain and France and, of course, Ireland, and elsewhere—in containers, supposedly, though their methods of disposal at Sellafield do not encourage me to imagine that their methods elsewhere should be assumed to be particularly cautious.

I suppose the British make lots of money cleaning spent fuel rods from all over the world, and from their own facilities. To be a source of a substance so prized as plutonium must bring wealth, and influence too. It is certain that they do not do it for their health. Exactly contrary to the universally held view, Britain is an island of unevolved laissez-faire plutocracy characterized by unregulated (my translation of the British "self-regulated") commerce and industry. So far from being lumbered with the costs of runaway socialist largesse, Britain ranks near the bottom in Europe not only in health spending but also in spending for education. In workers' wages and benefits, it has never approached the levels achieved by West Germany, Sweden, or the Netherlands. The British seem rather fond of their poverty, which I think is a social and economic strategy rather than the mysterious, intractable affliction it is presented as being. It effectively excuses the state from responsibility for the conditions of life of the poor, and for the quality of life of ordinary people. While lowering public expectations, this "poverty" justifies the astonishing recklessness of British industries, public and private, and makes it entirely acceptable for government and industry to be in cahoots to a degree that boggles the American mind.

Avoiding costs to industry is treated as an unquestioned good—Britain being so poor, after all. That very little trickles down from these coddled industries is a fact blamed squarely on the British worker, of all people, who, if he is lucky, toils for bad pay in a decaying factory and hopes that his children's lives will not be worse. Only consider: Britain is the world's fourth-largest arms dealer, a major exporter of petroleum, a major exporter of drugs and chemicals, a major center of banking and insurance, a major center of tourism. And it has access to the vast literatures of research and technology produced in the United States, the application of which in other countries is slowed and complicated by the problems of translation. This seems to me to be the basis for a presentable economy. But no, Britain is "poor"—because its workers are sullen and Luddite, or because its governing classes are too haplessly genteel and fair-minded to cope in the hurly-burly of the marketplace, or because the national character has grown idle in the embrace of the Welfare State, or because the great forces of entropy and decline have at last overtaken this noble civilization. Or because neither law nor custom encourages the sharing of wealth. Consider: University students are almost entirely subsidized. But only 5 to 7 percent of secondary-school students are admitted to universities. Since nothing is done to correct for the advantages children of privileged backgrounds bring to examinations and interviews—compensatory education is expensive, and Britain, after all, is poor—the subsidies go to the children of the prosperous. The cost per student of the university system to the state justifies its being kept very small—and this magnifies the value and the prestige that attach to university degrees. That is British socialism.

My point is simply that all the talk of decline, along with the continuous experience of austerity, creates an atmosphere in which the granting of enormous latitude to corporations, whether private or public, seems urgently necessary, and the encumbering of them with codes and restrictions a luxury embattled Britain can scarcely afford. Economic considerations have an importance and a pervasiveness that startle. The *Sunday Times*, in reporting the critical study of the British diet that had been suppressed, laid the blame to government fear of a negative impact on the food industry, and also to an awareness on the part of the government that old people are expensive: "Civil servants representing the social services . . . point out that healthy and long-lived citizens will increase the number of old-age pensions." Britain, you must always remember, is poor.

What a thoroughly miserable business. What arrogance to save a few quid by allowing Sellafield to spew and hemorrhage, again and again, on and on. According to the *Sunday Times*, a spokesman for British Nuclear Fuels agreed that it was "in everybody's interests to get discharges down as low as possible" but argued that the cost was "prohibitive." He said, "We would have to pass the cost on to our customers, which would mean higher electricity prices. We are already spending £500 million on reducing our discharges. We have reduced them considerably over the past ten years." Reduced them from what, *to* what? Note how "everybody's interests" are put in the scales against cost, and with what result. Why should expenses at a fuel reprocessing plant raise the price of electricity, rather than of plutonium? And why should the cost of recycling spent fuel for Japan—to pick a name out of the air—be subsidized by consumers in Britain? The idea is preposterous. We are hearing the same old song: *Shackle us with restrictions and you will pay dearly for it.*

Con Allday, chairman of BNF and, as one may glimpse him through the dark glass of British newspaper journalism, a man of views as emphatic as they are liable to be consequential, and who was quoted in the *Guardian* as saying that "there is little point in spending additional money simply to be safer than safe," is well deserving of some attention, while we are on the subject of thrift. This gentleman, according to the *Guardian*, "announced a new feasibility study into how the company can reduce radioactive discharges into the Irish Sea to 'as near zero as possible.'" I am quoting this so that you can share my admiration of the language. "He said: 'Public acceptability of nuclear power is so important and the time-scale needed for a swing-round of public opinion is so long that we must be realistic and accept that our discharges must be reduced to very much lower levels than hitherto planned.' [This was] 'even though there is no rational, cost-effective basis for doing so on risk assessment grounds.'" Weighing cost against risk again. That really is an interesting exercise, quite theological, I think. Considering that the expense involved in running a nuclear plant safely is truly vast, is it possible to say that the value of a given number of lives is exceeded in cash terms by the expenditure that would be required to prevent their loss? Clearly for these purposes the answer is yes, a fact all the more disturbing since the question is gravely skewed by the association of this slovenly enterprise with "nuclear power" and by the insistence—based on what?—that anyone, least of all an island of coal in a sea of oil, needs nuclear power in any case. Note Allday's impatience with the idea that discharge levels lower than Sellafield's should be achieved. Does this give us insight into the environmental standards maintained at other facilities?

Even Dr. Black, whose report found that the connection between radiation and leukemia at Sellafield was "by no means proven," was quoted by the *Guardian* as having said that "the risks of living near Sellafield were no greater than many of the risks everyone faced in their daily lives. He compared the increased risk to that of someone who used a private car rather than public transport." This unctuous little simile translates into an admission that there is some measurable risk involved in living near Sellafield. (Risk of what? Leukemia, surely, among other things. Then is not the presence of leukemia this very risk actualized? By no means proven!)

How has this happened? I can only speculate that within a tiny community of specialists, where esteem, advancement, and influence travel through a very narrow channel, and where over the life of a discipline the views of a very few people are reflected in policies of great magnitude and consequence, dissent would have little practical or emotional reward. Choices have been made, by scientists, industrialists, and politicians, that have reflected their willingness to accept human deaths at a certain rate, to put a part of the earth at risk, and the sea, contaminating them irreversibly. They have presumed so far on the basis of notions about the hazards involved that they admit to be conjectural. This is an appalling presumption, truly unpardonable if their notions prove wrong. It ought to be expected, therefore, that their standards of proof would be exceptionally rigorous.

Certainly the development of these policies has been very much affected by the dangers, political and diplomatic, of the issues involved. The British would know the effects of radioactivity if they had monitored the Australians who lived in the path of fallout from the huge, misbegotten hydrogen-bomb test at Monte Bello; or the aborigines who drifted across Maralinga, in South Australia, where radioactive detritus was left behind after British weapons testing; or the populations affected by the fire and the radioactive cloud that drifted southeast and west from Sellafield in 1957. They have given themselves many opportunities to look into this question and availed themselves of none of them, no doubt because to do so would undermine their claims that nothing serious has really happened.

There is, as I have said, the continuing threat of economic erosion to keep the public mind focused on the short-term and the local; and there is the image of the government battling to recoup Britain's losses and restore her scanted dignity; and there is the educational system, which trains very few people and these very narrowly, greatly enhancing the authority of specialists while diminishing the content and forcefulness of public debate and the numbers involved in it. And there is the secretiveness that permeates British life, which allows the Foreign Office to impound the records having to do with Argentina's claims to the Falklands; which prohibits journalists from reporting what they see in prisons; which conceals the identity of those on the committees that choose Britain's magistrates (the magistrates have no legal training—they simply suit some anonymous notion of worthiness); which leads the governing bodies of cities and counties to conduct their business behind closed doors. The Official Secrets Act is simply the most conspicuous manifestation of all this. Granting that it is used as the basis of prosecutions, and assuming that the *Guardian* is accurate in its accounts of mail openings, wiretappings, and housebreakings practiced by MI5 and MI6, the British secret police, against groups such as Greenpeace, Friends of the Earth, and the National Union of Mineworkers—nevertheless, it seems to me that the English, at least, have the government they deserve, that they prefer not to know, and that they have very little capacity for exerting power and influence. I think they feel—deeply feel—that their moral rectitude is preserved intact by this means. The Greenham Common women will never encircle Sellafield, though Britain could desist unilaterally from its war against the sea, which is not a terrifying threat, but a terrifying fact.

Then there is the absence of American reaction to consider—especially puzzling since both Greenpeace and Friends of the Earth have been involved with Sellafield. British Greenpeace was given a heavy fine—paid by public donations—for tampering with the pipeline and was induced to intervene to prevent Danish and Dutch Greenpeace ships from sending divers down by the threat that all its resources would be sequestered by the court if they did. Why Greenpeace has chosen not to galvanize public opinion outside the range of such restrictions, I cannot imagine. Perhaps regional patriotism has stood in the way of global matriotism. Or perhaps British environmentalists, like many Europeans of advanced views, believe that American public opinion is too brutish to be enlisted in any good cause. It is a treasured faith among Europeans of the right and the left that we are a nation of B-movie villains laying waste to the continent and to one another by any means that come to hand, a sort of frenzy of capitalist rapacity.

Europeans on the left enjoy the opinion that they are very advanced thinkers. In fact they are simply intellectual cargo-cultists, to whom accident now and then delivers an elaborated policy, a sophisticated idea, or half of one. That crude, capitalist America should enforce higher standards of public conduct than humane, socialist Europe is not to be imagined. So our example in environmental matters is almost never consulted, and our research and experience are almost never invoked.

We are greatly at fault in this. There is a streak of pure yokel that reaches straight to the top of American intellectual life, widening as it goes, and it is deference toward all things "English." We cannot believe that the English could be stupid or corrupt. We think of them as our better selves, and the source of our most precious institutions—a slander on the dark and the ethnic and a disparagement of the noisy public dramas of advocates and adversaries that provide us with the legal and ethical capacity for discrimination and judgment. We are capable of outrage and we are capable of shame, like a living soul. If we are fortunate in one thing it is in the knowledge that we *can* do evil, and we *can* do injury. A country incapable of scandal is like a mind incapable of guilt or a body incapable of pain.

On the twenty-fourth of July, the *Guardian* concluded its editorial on the Black report, titled "Lingering Particles of Unease," with a call for "one group of inter-disciplinary experts who do nothing else but shadow it round the clock." In the editorialist's affable view, "life with [Sellafield] is a tumultuous and ongoing affair." On the thirtieth of July, the *Guardian* wrote that Charles Haughey, the former Irish prime minister, had called the report a whitewash. He said: "If there is a high incidence of leukemia in an area where a nuclear plant is situated, surely to God the obvious interpretation is that the plant was responsible for it. These figures alone would in my view justify closing down the plant immediately for further investigation, and certainly putting a lot of people in gaol who have clearly been telling us lies over the past four or five years about this matter." The words we have longed to hear, but from the wrong side of the Irish Sea. May we still hope for decency, at long last?

BURYING NUCLEAR WASTE[3]

After nearly 30 years of politicking and scientific dispute, the Department of Energy (DOE) has announced the selection of three sites for the disposal of radioactive wastes from nuclear power stations and weapons plants. Ongoing studies of their suitability will cut the list to a single, permanent repository that is scheduled to begin receiving wastes by 1998.

But even before the announcement, a panel of geologists and hydrologists at a symposium on waste disposal questioned the 1998 deadline. "There's a reasonable chance that it won't be met," said John Robertson, a groundwater specialist with a company that consults for the DOE.

Isolating the wastes for a period measured on the geologic time scale presents such enormous difficulties that Priscilla Grew, a geologist and member of the California Public Utilities Commission, called the disposal effort "a geoscientist full-employment act."

To date, 24 million pounds of high-level wastes, consisting mostly of spent fuel rods from nuclear plants, are languishing in temporary storage facilities that are running out of space. The total is expected to swell to 100 million pounds by the turn of the century. This material will fill the permanent repository to capacity, so a second site will have to be chosen during the 1990s to make room for wastes that will be generated during the twenty-first century.

Scientists still don't agree on exactly how long the wastes should be isolated from the environment, but the DOE advocates 10,000 years—enough time for almost all of the isotopes to decay. But critics point out that iodine 129 and plutonium, major constituents of spent fuel rods, will remain dangerously radioactive for millions of years.

[3]Reprint of an article by Tom Yulsman, *Science Digest* staffwriter. Reprinted by permission from *Science Digest*, 93:16. Jl. '85. Copyright © 1985 by the author.

As now envisioned, the repository will be a four-mile-square chamber mined out of rock or salt 1,000 to 4,000 feet underground. The waste will be converted into relatively inert pellets and encased in 160,000 heavily shielded canisters. Total cost for two repositories: $25 to $30 billion, which will be paid by consumers in the form of higher electric bills.

Each of the potential sites presents problems. Yucca Mountain in Nevada, for example, is situated in the basin-and-range province, which, according to M. S. "Doug" Bedinger of the U.S. Geological Survey (USGS), is literally being pulled apart by geologic forces. As a result, the bottoms are dropping out of valleys, and mountains are thrusting skyward. The lava rock of the Hanford, Washington, site, says Donald White of the USGS, is under heavy stress, raising the possibility of "rock bursting" that could breach the underground cavern and allow wastes to leak into the nearby Columbia River. The third site, in Deaf Smith County, Texas, sits below the Ogallala aquifer, the nation's largest underground supply of fresh water and the principal source of irrigation for the southern high plains. This region produces 10 percent of the nation's beef and 85 percent of the world's sorghum seed. (Two other sites, one in Utah and the other in Mississippi, have been chosen as backups.)

The difficulties involved in evaluating the Yucca Mountain site are typical. Geologists will have to predict rates of uplift and erosion to determine how deep to bury the wastes: too deep, and groundwater contamination could become a problem; too shallow, and the wastes could be uncovered as uplift causes increased erosion. The geologic record is the only guide, but, says Bedinger, there is no assurance that the future will repeat the past.

Predicting the behavior of mountains is difficult enough; knowing whether groundwater will seep into the cavern and carry radioactive residues into the environment may be even more of a problem. If they are to know the rate at which water will seep through the rock, scientists must study how the rock's microstructure—its network of cracks and fissures—will react to the heat generated by the wastes. According to Paul Witherspoon, of the Lawrence Berkeley Laboratory, the temperature of the rock will rise to between 200 and 400 degrees Fahrenheit for up to 1,000 years. "Large-scale underground tests will be necessary to simulate these conditions," he says. For each type of rock found at the three sites, five to eight years of experiments will be necessary. "This is one of the most difficult problems I've ever seen in the field of hydrology," Witherspoon says. "I'm afraid we won't make 1998."

In response, William Purcell, who is in charge of site selection for the DOE, says, "It's a fairly tight date to meet, but we will do our best."

While the scientists struggle with the technical complexities of waste disposal, politicians and activists are trying to derail the DOE's plans. Some are forming unusual alliances. The state of Texas, not known as a bastion of environmental sentiment, has joined forces with the Sierra Club in a suit against the DOE. Governor Mark White has vowed that he will fight, saying, "Sparks will fly before the people of Deaf Smith County will glow." And Governor Richard Bryan of Nevada has expressed similar sentiments, saying, "We don't want to replace the neon glow of Las Vegas with a radioactive glow."

Meanwhile, officials in San Juan County, Utah, are upset that they *weren't* chosen to host the world's first nuclear dump. Joe Slade, chairman of the San Juan County Travel Council, was quoted in a recent UPI dispatch as saying that a nuclear waste site would give southeastern Utah a "shot in the arm" by attracting more people to the area.

LOW-LEVEL LOWDOWN[4]

Thirty-five miles north of New York City, Union Carbide Corporation operates a chemical reprocessing plant and a small nuclear reactor. The facility produces some materials used for medical purposes and others used in making nuclear weapons. It also produces a liquid that is considered high-level radioactive waste under one section's regulations. But when the waste is put into 55-gallon barrels with only slightly contaminated paper and metal, the average concentration of radioactivity drops, and the resultant mixture fits within the commission's definition of low-

[4]Reprint of an article by Gale Warner, free-lance writer on environmental topics. Reprinted by permission from *Sierra*, 70: 19-23. Jl./Ag. '85. Copyright © 1985 by *Sierra*.

level radioactive waste. The barrels then make the 500-mile journey to the low-level waste dump near Barnwell, S.C., where they are placed in one of 32 trenches and covered with dirt.

While the Barnwell dump is considered a shining star in the waste-management business, significant levels of tritium, a suspected carcinogen, have been detected 200 feet southwest of its trenches.

This story is played out hundreds of times a year. The names and places change, but the problems of low-level radioactive waste stay the same. No one knows how to define it, who should be responsible for it, or how and where it should be discarded.

Currently defined by exclusion, low-level waste is that which does not fall into the high-level category, which includes spent nuclear fuel, reprocessing wastes, and mill tailings from uranium mining. What remains is a hodgepodge of trash from nuclear power plants, factories, research institutions, and hospitals— everything from laboratory animal carcasses and irradiated reactor components to emergency exit signs and residues from the manufacture of luminous watches. Much of this waste is relatively harmless. But some, particularly wastes from radiopharmaceutical companies and contaminated filters from nuclear reactor cooling systems, remain extremely potent for a long time.

Until the 1960s, disposal of low-level waste posed little difficulty. It was simply put into 55-gallon drums, loaded onto Navy ships, hauled out to sea, and unceremoniously dumped overboard. At 6,000 feet below sea level, experts believed, it would never be heard from again. (But it was, of course; the EPA has since found elevated levels of radioactivity in the seabed and marine life near California's Farallon Islands, where 47,000 barrels of low-level waste were jettisoned.)

With the increase in waste caused by the arrival of commercial nuclear power, it became more viable economically to replace ocean dumping with landfills, and to go from federal to private control. Commercial landfills were patterned after those at a dozen defense installations. Wastes were packaged in steel drums or wooden boxes and dumped into trenches that, when full, were covered with earth.

Apart from their design, low-level waste dumps went corporate with virtually no comprehensive planning or federal oversight. The first commercial dump opened near Beatty, Nev., in 1962. Maxey Flats, Ky., and West Valley, N.Y., opened in 1963;

Richland, Wash., in 1965; Sheffield, Ill., in 1967; and Barnwell, S.C., in 1971.

Three of the dumps were short-lived. West Valley closed in 1975, Maxey Flats in 1977, and Sheffield in 1978. Each facility stopped operating because radioactive materials had migrated off the sites. "There isn't a radioactive landfill in any area with 30 to 40 inches of rainfall a year that hasn't leaked," says Marvin Resnikoff, co-director of the Sierra Club Radioactive Waste Campaign. "Landfills act a lot like teabags: The water goes in, the flavor goes out."

But closure has not meant the end of problems at the dumpsites. At Maxey Flats, the largest of the closed dumps, plutonium has been detected more than a mile from the site. Groundwater contaminated with tritium continues to move out of the Sheffield site at the rate of a half mile per year. At West Valley, trenches have been infiltrated with water, created a "bathtub effect" that has spilled tritium and strontium into nearby streams.

The three remaining commercial dumpsites in Richland, Barnwell, and Beatty have handled all the nation's low-level waste for the last seven years. Residents and state officials are now getting tired of the situation. In October 1979 the governors of Washington and Nevada temporarily shut down the Richland and Beatty sites—the former because waste kept arriving "improperly packaged," in some cases with liquid oozing from the barrels, and the latter because of safety violations.

The governor of South Carolina, fearing that his state would become the low-level dumpsite for the entire country, immediately announced restrictions on the amount of waste the state would accept. Utilities and other producers of low-level waste protested that soon they would have nowhere to send their radioactive trash. All three states' actions were later rescinded, but they had created a crisis atmosphere that prompted Congress to pass the Low-Level Radioactive Waste Policy Act in December 1980.

Although several studies had recommended that low-level burial sites be returned to federal jurisdiction, the act dumped the problem into the laps of the states. It called on them to form interstate compacts and construct regional burial sites to handle each region's waste. Once negotiated, compacts were to be submitted to Congress for ratification.

The carrot—and the stick—of the bill was a clause allowing regions with ratified compacts to refuse to accept low-level waste from outside the region after January 1, 1986. The theory was that this would motivate states to reach agreement swiftly and begin constructing new dumpsites. Members of Congress patted themselves on the back for having permanently solved the problem of low-level waste while promoting interstate cooperation.

Putting the law into practice, however, has not been quite that easy. Some states have formed compacts, particularly those that now have dumpsites and are anxious to take advantage of the 1986 deadline. But others have delayed, hoping to leave the issue to political successors. In some areas negotiations have broken down or are in limbo as individual states flirt with several different regions, looking for the best deal. And even in regions that have formed compacts, new sites will not be ready until well past the 1986 deadline.

The Northeast, which generates 37 percent of the volume and 57 percent of the radioactivity of the nation's low-level waste, is in the worst shape. Negotiations for a Northeast compact fell apart when the three largest waste-generating states, New York, Pennsylvania, and Massachusetts, refused to join, leaving four smaller states that had already ratified the compact— Connecticut, Delaware, Maryland, and New Jersey—in the lurch. "They thought some big state would join and take the site, and now they don't know what to do," says Priscilla Chapman of the Sierra Club's New England Chapter. The comparatively tiny waste-generating states of Maine, New Hampshire, Vermont, and Rhode Island are adopting a wait-and-see attitude.

Massachusetts is in a unique position. In 1982 voters passed by a two-to-one margin a referendum requiring that any low-level waste facility not exclusively set aside for medical and institutional wastes be approved by the voters.

Meanwhile, the 24 states that do belong to regional compacts are waiting for congressional approval. It may be a long wait. Few members of Congress are likely to vote for anything that might cut off their state's access to the three current dumpsites.

In an attempt to bring order out of chaos, Rep. Morris Udall (D-Ariz.) has introduced a bill amending the 1980 act. The Udall bill recognizes that the 1986 deadline is unrealistic. At the same time it attempts to appease the states that have formed compacts and prod other states into action. The bill would postpone until 1993 the date that regions can refuse nonregional waste, but in the meantime it specifies reduced volume allocations for these sites.

"The Udall bill is on the right track," says Sierra Club Washington lobbyist Brooks Yeager, "but we don't think it goes far enough in addressing some of the key problems." Among these is the present definition of low-level waste, which lumps together wastes that are hazardous for a few years with those that must be isolated for hundreds or thousands of years. The Sierra Club believes the latter should be taken out of the low-level waste category and considered high-level, or perhaps put into a new category of intermediate waste. "Of course, just creating a new category and a new label isn't going to solve the problem," says Chapman. "You still have to figure out what to do with the wastes. But at least it recognizes that we can't just throw them all into one trench."

Still another question is whether shallow landfills are the most appropriate method for disposing of low-level waste. While the Nuclear Regulatory Commission remains committed to using trenches, others are not so sure. The states of Illinois and Kentucky, for example, will not make the same mistake twice; their two-state compact calls for "above-ground facilities and other disposal technologies providing greater and safer confinement."

Segregating wastes at the source and storing them in above-ground facilities significantly lessens the amount of low-level waste to be dealt with, since wastes that decay to harmless levels of radioactivity in just a few years could then be disposed of as regular trash. Wastes with hazardous lives longer than the expected lifespan of a facility could be removed and sent to a high-level waste repository when one becomes available.

While the initial costs of above-ground storage facilities are higher than the cost of simply digging a ditch, they have several advantages: Waste can be easily monitored and leaking packages identified; the costs of pumping out leaky landfills are eliminated; and storage can be located in any part of the country because siting is less dependent on climate and geology. Logical places to put above-ground facilities might be the sites of defunct nuclear reactors, which "are going to be de facto waste sites for some time to come," says Resnikoff.

The utilities, however, have been reluctant to take any role in either temporary or long-term storage of low-level waste, says Yeager. "They get upset any time someone asks them to take responsibility for the waste they produce."

"Waste disposal is already very expensive, and our concern is that utilities will have to increase rates, hospitals will have to increase fees, and universities will have to cut back on their research," says Mary Paris of the New York State Low-Level Waste Group, a coalition of utilities, hospitals, universities, and manufacturers. "We could do above-ground storage, but it's a question of how much that would cost over the long term," she says.

According to Resnikoff, this kind of argument simply "runs interference" for the utilities, whose current expenses for low-level waste disposal are "so low that you couldn't measure it on your electricity bill." Medical wastes account for 7 percent of the volume and less than 1 percent of the radioactivity of the nation's low-level waste stream; in contrast, utilities produce 54 percent of its volume and 24 percent of its radioactivity. Medical wastes contain mainly short-lived isotopes that can be safely stored in a warehouse for several years and then disposed of as regular trash.

Meanwhile, South Carolina is threatening to close its dump if Congress doesn't stop dragging its feet on ratification of the Southeast compact. Should the Barnwell site actually close, "it would create quite a problem," says Yeager. "The utilities really hold the key to the short-term capacity crisis. It would cost them money, but they can afford to store it and they have the room."

Overall, Resnikoff is encouraged by what he sees as a movement away from landfills, and the opportunity presented by the Udall bill to redefine low-level waste. However, it is crucial that the public be informed if the problems are to be resolved safely. "If we weren't around looking over industry's and regulators' shoulders, my guess is that all this stuff would just be tossed into the backyard," says Resnikoff. "Only with a tremendous amount of citizen activity on this issue can we continue to make headway."

THE DEADLIEST GARBAGE OF ALL[5]

For Congress, year after year, Christmas is the season to commit folly. So it was last December 19 when the holiday-rushed lawmakers tried to repair a 1980 error called the Low-Level Radioactive Waste Policy Act. Hours before adjourning, the members heeded dire warnings that inaction would not only shut down the nation's nuclear-waste dumps by New Year's Day but also halt lifesaving medical research throughout the land. Duty outweighed deliberation. Congress quickly approved various, supposedly beneficial amendments—thus blotting the books with yet another multiplier of the country's nuclear-waste problem.

It was virtually a replay of the "crisis" that spawned the original flawed law just before (yes) Christmas 1980. Unless that measure passed, Arizona congressman Morris Udall warned back then, the country faced "the prospect of our hospitals and research institutions, as well as commercial nuclear power facilities, being forced to shut down." Blanching at that prospect, Congress hurriedly backed Udall and enjoyed a guilt-free holiday recess.

The 1980 law declared that each state—not the federal government—would now be responsible for dealing with its own low-level nuclear waste. The deadline for compliance was January 1, 1986.

Because the deadline was totally unrealistic and the law provided little or no incentive to meet it, the 1985 Christmas "crisis" produced a seven-year extension, postponing final compliance until January 1, 1993. Along the way, the states are obliged to meet interim goals and submit regular progress reports to Congress.

In effect, the revised law pressures the states to comply by creating *more* low-level nuclear-waste sites, not fewer, and to speed up this proliferation at the expense of better waste management. Getting the job done overshadows how well it's done.

[5]Reprint of an article by Susan Q. Stranahan, member of the editorial board of the Philadelphia *Inquirer*. Reprinted by permission from *Science Digest*, 94:64, 67–68, 80–81. Ap. '86. Copyright © 1986 by the author.

Instead of the current three low-level nuclear-waste sites, located in remote areas geologically suited for radiation containment, the nation will end up with 12 or more, some of them unavoidably located in populous, geologically unsuited areas. And because of the tight timetables in the 1985 amendments—plus regulatory inertia—the states will be discouraged from using innovative waste-disposal technologies. Moreover, if their sites aren't operational by the 1993 deadline, they could be subject to huge damage suits by private companies producing nuclear waste. This makes it all the more probable that many states will stress appearances and speed rather than technological excellence.

Not only has Congress failed to solve the waste-management problem, it has apparently made it worse.

America now generates some 2.6 million cubic feet of low-level nuclear waste each year, containing 600,000 curies. A curie is the standard measure of radioactivity—technically, the amount of a substance required to create 37 billion atomic disintegrations per second. Those disintegrations produce the energy, or activity, of a radioactive source, and that activity lessens, or decays, over a period of time. Each radioactive substance has its own characteristics. Some may decay to the point of harmlessness in less than a day; others emit energy and remain hazardous for thousands of years. Some pose a threat to humans by mere external exposure; others present a danger only if the substance is inhaled or swallowed.

In recent years, the scientific debate over how much radiation is harmful has intensified. While some authorities, including those setting federal exposure standards, believe that the risks below a certain level of radiation each year are imperceptible, others argue that exposure to any amount may cause damage. It's worth noting that every time the federal government has revised exposure-safety guidelines since the first ones were issued by an international committee in 1934, it has lowered them. Once the allowable exposure rate was 73 rems per year (a rem is a standard measure of radiation dosage); it is currently 5 rems per year for occupational exposure—a level some scientists believe still exposes countless Americans to undue risks of cancer and genetic defects.

Almost 99 percent of the curies in commercial low-level radioactive waste in this country comes from nuclear reactors and industries; the remainder, from medical and scientific research. Low-level waste is defined as what it is not. It is *not* uranium-mill tailings or irradiated nuclear fuel. Everything else—such as laboratory gloves slightly contaminated with technetium 99m, which loses half its radioactivity every six hours, or resins and sludges contaminated with cesium 137, which loses half its radioactivity every 30 years—constitutes low-level waste. While much of this debris is truly low in radiation levels, some remains hazardous for hundreds of years, and even brief exposure to it could be fatal.

Another way to look at the inherent health risks posed by low-level waste is to consider that it consists of some of the same materials—cesium and cobalt, for example—that were present in fallout from nuclear-weapons tests conducted in the 1950s and early 1960s. Fallout has been blamed by some groups for causing numerous cancers among those who lived downwind of the Nevada test site. In a recent court trial, nine people were awarded compensation for cancers they claim were caused by fallout. Although the federal government maintained that the amount of radiation in the fallout was not dangerous, a federal court judge weighed vast quantities of scientific evidence and concluded otherwise. If it survives appeal, the 1984 decision by Judge Bruce Jenkins of Utah will stand as a landmark in linking low levels of radiation exposure to damaging health effects.

Easiest and Cheapest

Historically, the federal government and those who generate radioactive waste made no distinction among the types of low-level waste, disposing of it all the same way: by burying it in containers in shallow trenches in the ground. That was the easiest and cheapest way to get rid of it. Whether or not it was the safest was another matter. As long as the general public remained convinced by government assurances that burying the debris was perfectly acceptable, federal officials adopted the policy that management of low-level waste was a private-sector enterprise.

They had another reason to keep quiet. The government itself produces large quantities of nuclear waste and wanted to minimize its disposal problems. Much of that waste is high-level and extremely dangerous. Plutonium, a by-product of nuclear-weapons fabrication, is one of the deadliest substances produced by man. It loses only half its radioactivity in 24,000 years, and a microscopic speck can cause cancer and death. Yet, the government didn't classify it as high-level waste, which made disposal a lot simpler.

Pounds of plutonium, produced by the government and, for a while, by private nuclear-fuel processors, were dumped in trenches in designated facilities until 1970. The Atomic Energy Commission then decided to halt the practice—but only on federal land. It imposed no similar ban on private property. As a result, plutonium was dumped in all but one of the commercial low-level nuclear-waste facilities. Some of it came from the U.S. government, which evidently regarded the material as too hazardous for *its* property, but not for someone else's. The practice finally ended in 1979, leaving a still vivid memory of the government's less than responsible attitude toward radiation hazards.

The federal government is now the sole U.S. producer of plutonium, which has been buried or stored at a number of sites around the country, including three vast Department of Energy (DOE) installations in Washington State, Idaho and South Carolina.

The United States once had six privately run low-level radioactive-waste burial grounds. Today it has three: in Barnwell, South Carolina; Hanford, Washington; and Beatty, Nevada. The others—at Maxey Flats, Kentucky; West Valley, New York; and Sheffield, Illinois—were closed between 1975 and 1978.

At the three closed facilities, trenches have leaked and radioactivity has spread to adjacent soil. Plutonium has been detected moving from trenches at West Valley and Maxey Flats. The three operating graveyards have not been without their problems, either. Radioactive tritium was discovered leaking from trenches at the South Carolina site. A fire in Nevada, caused by improperly stabilized uranium, exposed 10 people to radiation. Improperly packaged waste has arrived at the Nevada and Washington sites in containers that had split open and spilled their contents somewhere along the way.

No new burial facilities have opened since 1971, when Chem-Nuclear Systems Inc. began operations at Barnwell, South Carolina. Yet the output of low-level radioactive debris grew throughout the 1970s. Although the federal government insisted that devising a low-level-waste management program was its responsibility—in part to avoid an "undisciplined proliferation" of new dumps all over the country—it did nothing. Antinuclear sentiment had grown in the nation, and nobody in Washington was eager to take on that political tarbaby.

The Governor Protests

Things came to a head in 1979. The governors of Nevada and Washington temporarily closed their nuclear burial grounds, and South Carolina began receiving 80 percent of the nation's atomic garbage. The state's newly elected governor, Richard Riley, protested, declaring that his state would no longer serve as the "solution" to the nation's nuclear-waste dilemma—a major policy reversal, because South Carolina for years had actively courted nuclear industries, apparently willing also to accept their waste.

Governor Riley cut by half the amount of nuclear trash that Barnwell could accept over the next two years and indicated he might close the state-licensed facility or impose even sharper volume limits. By threatening to create a national disposal crisis, he forced Congress to act. It did so, but hardly in a deliberative manner, choosing what appeared to be the easiest and cheapest technical option.

With a smattering of debate, the 1980 Low-Level Radioactive Waste Policy Act was approved. By January 1, 1986, states would become responsible for their own waste. (The law did not affect waste generated at federal installations; responsibility for that remained with the DOE.) The law envisioned the formation of regional compacts among states and the development of one or more burial grounds within each compact. States that did not join a compact could be excluded from using the graveyards in other regions.

Unfortunately, the law passed by Congress was designed more to make states share the political burden of nuclear waste than to improve its disposal. As a result, proliferation is inevitable. The law's time constraints, for example, pressured states to continue relying on landfill disposal, a practice that has failed almost everywhere it has been used, especially in areas with any appreciable amount of rain. Although Congress has now given the states more time for compliance, many observers say nothing has changed. Expediency still rules.

The facts support that view. States that want to use engineered containment, such as aboveground concrete silos for the waste, face major obstacles. The Nuclear Regulatory Commission

(NRC), which will set the standards for disposal and approve construction plans, has no applicable regulations for that technology. The General Accounting Office estimated it would take a minimum of five years to site, license and build a conventional burial facility; it made no predictions about how long it would take if alternative technologies were used.

Another issue that may get pushed aside is the actual definition of low-level waste. Last year, Pennsylvania officials urged Congress to reclassify the small volume of low-level waste that remains radioactive for hundreds of years. These long-lived, more concentrated materials, called Class C waste, constitute only one or 2 percent of the volume but represent 95 percent of the total curies produced each year in the commercial sector. They include worn-out reactor parts and moderately radioactive resins and sludges.

Pennsylvania officials argued that if the federal government assumed responsibility for Class C waste—as it has agreed to do for utilities' reactor fuel—the safety problems of low-level-waste sites would be greatly reduced. That effort failed, in part because of opposition from officials in South Carolina, Nevada and Washington who argued, in essence, "If we had to take it, everybody else should, too." Also opposed were the NRC and the DOE, the latter because it would have had to accept Class C waste at its facilities.

Meantime, the law doesn't preclude states themselves from segregating the Class C waste or from selecting sophisticated containment systems. It simply provides no incentives for them to do so.

Jobs vs. Radiation

Another paradox in the current waste debate becomes apparent on a visit to Barnwell, population 5,600. Buried in shallow trenches a few miles outside town is something nobody else wants: 17 million cubic feet of low-level nuclear garbage. Each year, that volume grows by 1.2 million cubic feet. Yet, when asked about the Chem-Nuclear facility, most people on the street speak of jobs and tax revenues, not radiation. They are the beneficiaries of a 20-year campaign by state officials, which ended with Riley's election, to turn South Carolina into what has been described as a "nuclear enterprise zone." The state has so many nuclear-related industries that in 1984 it led the nation in low-level-waste production—255,000 cubic feet. Many Barnwell residents seem to be as pleased by all this as they are perplexed by the governor's threat to close the burial ground or sharply reduce its operations at the end of 1992, when other states are supposed to assume responsibility for their waste. "Logic is being thrown to the wind," said Steve Wright, executive director of the Barnwell County Chamber of Commerce and Industrial Development Commission.

Chem-Nuclear isn't Barnwell's only nuclear neighbor. On the other side of town sits a giant that makes the small atomic graveyard seem almost benign. The massive Savannah River Plant, the federal government's weapons-grade plutonium production facility, occupies 300 square miles between Barnwell and Aiken. The 35-year-old plant is the state's largest single employer with over 10,000 people on the payroll. More than a cumulative total of 30 million gallons of high-level nuclear waste have been stored "temporarily" for decades in underground tanks, leading many South Carolinians to believe that their state has become a de facto high-level-waste repository. Airborne radioactive emissions are commonplace, and an unwritten rule of thumb at the plant is that nuclear pollution is acceptable if it doesn't go off site.

"Why should I worry about a little tritium leaking from Chem-Nuclear when one million curies of tritium and krypton 85 escape into the air every year from Savannah River?" asks Michael Lowe, head of the antinuclear Palmetto Alliance, which nevertheless supports Riley's efforts to shut the Chem-Nuclear facility. "By closing down Barnwell and creating a crisis," he says, "you will force solutions."

But are they the right solutions?

South Carolina, Washington and Nevada have succeeded in what they wanted to do, which was to spread the responsibility for managing low-level nuclear waste more equitably among those states that produce it. They achieved their goal by convincing Congress that a solution to the problem was simple and could be implemented in just a few years.

What Congress produced instead—a nonsolution—was clearly unclear to many lawmakers as the hour grew late last December 19. "This is a great day for America," declared South Carolina senator Strom Thurmond. "This is a wise step the Congress is taking." On the House side, the response was somewhat less grandiose. Seconds after the amendments were rushed through, the real issue was raised by Minority Leader Robert Michel of Illinois: "Mr. Speaker," he asked, "where are we going and, more importantly, when are we quitting?"

THE CLOSING ACT:
DECOMMISSIONING NUCLEAR POWER PLANTS[6]

Nearly four decades and 350 power plants into the nuclear age, the question of how to dispose safely and economically of nuclear reactors, in addition to their wastes—is largely unanswered. Unlike other electric generating technologies, nuclear plants cannot simply be abandoned at the end of their operating lives or demolished with a wrecking ball. Radioactivity builds up each year the plant operates, and all of the contaminated parts and equipment must be securely isolated. Some radioactive elements decay quickly, but others remain hazardous for millennia.

No one knows how much it will cost to decommission the hundreds of units in service or under construction around the world. Estimates range from $50 million to $3 billion per plant. The enormous rush before 1980 to build nuclear power plants means that much of the nuclear decommissioning bill may fall due in a single decade—from 2000 to 2010—to be paid by a generation that did not take part in the decision to build the first round of nuclear power plants and that did not use much of the power generated. Although engineers still debate about the life expectancy of a reactor, economical operation may not be feasible for longer than thirty years. Numerous technical difficulties, including the constraints radiation build-up places on routine maintenance and the inevitable embrittlement of the reactor pressure vessel, are likely to limit opportunities to extend plant life.

The oldest commercial nuclear reactors are already nearing the end of their useful lives, and some plants have closed prematurely because of accidents or faulty designs. In the United States,

[6]Reprint of an article by Cynthia Pollock, researcher at Worldwatch Institute for energy policy and technology issues. Reprinted by permission from *Environment*, 28:10–15, 33–36. Mr. '86. Reprinted with permission of the Helen Dwight Reid Educational Foundation. Published by Heldref Publications, 4000 Albemarle St., N.W., Washington, D.C. 20016. Copyright © 1986.

dozens of tiny research and military reactors are no longer used, four small commercial units are shut down and awaiting decommissioning, and the U.S. Nuclear Regulatory Commission (NRC) estimates that another 67 large commercial units will cease operations before the year 2010. Worldwide more than 20 power reactors are already shut down; 63 more will probably be retired by the turn of the century and another 162 between 2000 and 2010. Countries with advanced nuclear programs will soon start to feel the pressure associated with managing the new and broadened back end of the nuclear fuel cycle that includes low-level radioactive waste from decommissioning as well as the high-level radioactive waste of spent fuel.

Decontamination, Dismantlement

Decommissioning is waste management on a new scale, in terms of both complexity and cost. Following plant closure, the company or agency responsible must first decide which of three courses to follow:

• decontaminate and dismantle the facility immediately after shutdown;

• put it in "storage" to undergo radioactive decay for fifty to one hundred years before dismantling; or

• simply erect a "permanent" tomb.

Each option involves shutting down the plant, removing the spent fuel from the reactor core, draining all liquids, and flushing the pipes. Elaborate safeguards to protect public and worker health must be provided every step of the way.

Under the immediate dismantlement scenario, irradiated structures would be partially decontaminated, radioactive steel and concrete disassembled using advanced scoring and cutting techniques, and all radioactive debris shipped to a waste-burial facility. The plant site would then theoretically be available for "unrestricted" use.

Plants to be mothballed, on the other hand, would undergo preliminary cleanup, but the structure would remain intact and be placed under constant guard to prevent public access. After fifty years most of the short-lived radioisotopes would have decayed, further safety gains would be negligible, and the facility would be dismantled.

Entombment, the third option, would involve covering the reactor with reinforced concrete and erecting barriers to keep out intruders. Although once viewed as the cheap and easy way out, entombment is no longer considered a realistic option because of the longevity of several radioisotopes.

A survey of 30 electric utilities in the United States revealed that 73 percent planned to dismantle and remove their reactors promptly following shutdown. In contrast, utilities in Canada, France, and West Germany are planning to mothball most of their reactors for several decades before dismantlement.

Regardless of the method chosen, decommissioning a large nuclear power plant is a complex engineering task, without precedent. The high levels of radioactivity present at recently closed reactors place numerous constraints on the decommissioning crew. Workers must take elaborate precautions and limit their time in contaminated environments. Radiation exposure must be carefully monitored, and adhering to regulations can greatly reduce shift length. Worker productivity is unavoidably low, less than half of what it could be in a nonradioactive environment.

Much of the radioactivity in a retired nuclear plant is bound to the surface of structural components. The type of material and its exterior surface determines the depth of penetration. The range is typically from as little as several millimeters to as much as 15 centimeters for unsealed concrete. Although some surface contamination can be washed off by using high-pressure water jets and chemical decontaminants, only a fraction of the material becomes clean enough to recycle or dispose of in commercial landfills.

The volume of solvents used in decontaminating surfaces must be carefully regulated because the effluent also becomes radioactive. Spills during either operation or cleanup can result in contamination of the surrounding soil. Keeping waste volumes to a minimum is an elusive goal: each piece of machinery and every tool that comes into direct contact with a contaminated surface must be decontaminated or added to the radioactive waste pile.

In addition to contaminated structural waste, "activation" products are the other source of radiation confronting decommissioning crews. When nuclear fuel undergoes fission—the splitting of uranium atoms—stray neutrons and other particles escape and bombard the nuclei of atoms in the surrounding structures, and the resulting change in composition causes some elements in the steel and concrete that encircle the reactor core to become radioactive.

For the several decades following plant shutdown, the most problematical elements are those that decay the fastest. Measured in curies, or disintegrations per second, cobalt and cesium are the dominant short-lived radioisotopes in contaminated materials. Other elements with longer half-lives (the time it takes radioisotopes to decay to half their original levels) are present in smaller quantities and will dominate radiation levels in the future. Significant amounts of long-lived nickel and niobium radioisotopes are present in neutron-activated wastes and will probably render the wastes unsuitable for traditional shallow-land disposal. The longest-lived hazardous element detected to date, nickel-59, has a half-life of 80,000 years. Overall, neutron-activated components contain over one thousand times the radioactivity of contaminated components.

Following preliminary decontamination, the reactor and surrounding structures must be dismantled into smaller pieces for transportation and burial. A pressure vessel containing a 1,000-megawatt reactor is typically over 12 meters high and 4 meters in diameter and may not be able to be shipped intact. But cutting it into pieces is hazardous and expensive. Each cut causes more airborne contamination and greater worker exposure. Remote-control operations and the need to keep dust formation to a minimum complicate the dismantlement.

Decommissioning just one large reactor would yield a volume of contaminated concrete and steel equal to one-sixth of the low-level radioactive wastes now produced in the United States each year. Decommissioning all U.S. operating reactors would yield well over 1 million cubic meters of low-level waste, enough to build a radioactive wall 3 meters high and 1 meter wide from Washington, D. C., to New York City. Some of the wastes are considered too radioactive for shallow land burial. NRC has identified "several kinds of decommissioning wastes for which disposal capacity is presently either not available or not assured under the current statutory and/or regulatory framework."

New, larger waste containers will minimize the extent of dismantling required and volume-reduction techniques now being developed may substantially reduce the number of waste disposal shipments. One of the most obvious and simplest approaches to reducing waste volume is to melt the steel. Other strategies include nesting the components and compacting them under high pressure.

Decommissioning Track Record

Practical decommissioning experience has been limited to very small reactors. The 22-megawatt Elk River plant in Minnesota is the largest that has been fully decontaminated and dismantled. The U.S. Department of Energy (DOE) completed the three-year project in 1974 at a cost of $6.15 million. Underwater plasma arc torches cut apart the reactor and 2,600 cubic meters of radioactive waste were disposed of at government burial sites. Today's reactors can produce fifty times more power and will have operated for more than seven times as long as the Elk River reactor. Since radioactivity builds up in proportion to plant size and operating life, a 1,000-megawatt nuclear reactor used for thirty years would be considerably more contaminated.

Twenty-five miles outside Pittsburgh, Pennsylvania, the small 72-megawatt Shippingport reactor is currently being decommissioned by DOE. The plant, opened in 1957 as the first U.S. commercial power reactor, has had three different cores and has already undergone one round of decontamination. DOE plans to encase the 10-meter-high steel reactor vessel in concrete, lift the 770-ton behemoth intact, and send it by barge down the Ohio and Mississippi rivers, through the Gulf of Mexico and the Panama Canal, and up the Pacific coast and the Columbia River. It will finally be buried in an earthen trench on the government-run Hanford nuclear reservation in south-central Washington state.

Keeping the Shippingport reactor pressure vessel in one piece will lop at least $7 million (7 percent) off the total cost of decommissioning it, but this action is shortsighted. Other reactors may be too big to ship in one piece, and the most difficult task decommissioning crews will face is dismantling the pressure vessel and its contents. Tackling that problem now at Shippingport could provide valuable knowledge and experience.

In Europe, plans for dismantlement of several commercial reactors are currently being considered. The first three projects are the 100-megawatt Niederaichbach unit in West Germany, the 33-megawatt Windscale advanced gas reactor in the United Kingdom, and the 45-megawatt Marcoule G-2 gas reactor in France. Although the French and U.K. plants are small, each was in operation for about twenty years. The larger German reactor was in service for two years when technical difficulties closed it down. Each unit has a different design and problems unique to specific technologies are likely to be discovered.

Finally, experience gained at Three Mile Island—the site of the most serious nuclear power mishap, where the reactor core partially melted—will also aid future decommissioning work. The industry's knowledge of robotics, chemical decontaminants, and remote cutting techniques has greatly expanded as a result of the cleanup. The level of contamination at Three Mile Island is many times higher than will be encountered at most power reactors; cleanup costs there are projected to pass $1 billion before the process of decommissioning is contemplated.

In selecting a decommissioning schedule and the appropriate decontamination and dismantlement methods, the overriding consideration must be worker and public safety. Although radioactivity declines more than tenfold during the first fifty years after plant closure, thereby reducing worker exposure, the reactor is a potential hazard to more people during the cool-down period. The retired Humboldt Bay reactor on the northwest coast of California is a good example. It lies in a seismically active zone and is not structurally equipped to handle tremors—which is why it was permanently taken out of operation in 1976. Immediate dismantlement would make the site available for other uses and would limit potential public exposure to radiation. Yet efforts to dismantle the plant are not expected until after the turn of the century.

Estimating Cost

The cost of decommissioning nuclear power plants is highly speculative. As noted earlier, the figures put forward run from $50 million to $3 billion per reactor. The majority of estimates cluster at the low end of the range. Some cost projections were adopted from generic estimates, others were based on a fixed percentage of construction costs, and a few were arrived at by using detailed, site-specific engineering studies. In effect, all figures put forward are guesses based on a number of assumptions.

In 1978 NRC asked the Battelle Pacific Northwest Laboratory to estimate the cost of decommissioning generic 1,100-megawatt pressurized and boiling water reactors (PWRs and

BWRs). (These designs account for 72 percent of all operating reactors worldwide—165 PWRs and 77 BWRs.) Battelle's figures ranged from $61.5 million to $86 million, depending on the plant-specific technology and the number of years after shutdown that dismantlement would be deferred. Immediate dismantlement of a PWR was estimated to be the least expensive, while waiting thirty years to dismantle a BWR was considered to be the most costly. In general, it is more expensive to dismantle a BWR, all other factors being equal, because of the greater volume of contaminated wastes that is produced.

For many years the lack of detailed plant-specific cost estimates led regulatory agencies and utilities to apply the Battelle figures to a great variety of facilities. Little notice was paid to differences in plant size and design, future availability of and distance from waste disposal facilities, and unique site characteristics such as space limitations or difficult topography. But as individual utilities began to conduct their own site-specific cost estimates and as various component costs (such as waste disposal) rose much faster than anticipated, it became obvious that the initial estimates were too low.

In 1984 Battelle updated its estimates for the Electric Power Research Institute (EPRI) and found that costs had indeed risen much faster than inflation over the preceding six years. Waste disposal costs rose the fastest. Assumptions were also modified to reflect current regulations and market conditions. The price tag for immediate dismantlement rose 69 percent for a PWR and 108 percent for a BWR. Comparative site-specific, rather than generic, estimates by another EPRI contractor for the same size plants resulted in estimates of $140.5 million for a PWR (35 percent higher than the updated Battelle estimate) and $133.6 million for a BWR, excluding the costs of removing nonradioactive structures.

Cost estimating guidelines recently developed by the Atomic Industrial Forum, an industry trade group, led members to project a decommissioning cost of $170 million per unit. This is more than a sixfold increase in ten years. (A report issued by the group in 1976 estimated that a 1,100-megawatt PWR could be decommissioned for $27 million.) And in Switzerland, a detailed three-year study of decommissioning costs concluded that retiring any nuclear plant would cost one-fifth as much as the facility originally cost to build. This translates into several hundred million dollars for recently completed plants.

Economist Duane Chapman, an independent analyst at Cornell University, predicts that decommissioning will cost as much as the original construction, in constant dollars. In the United States, this amounts to an average of about $3 billion for a new 1,000-megawatt unit. Chapman points to the complex procedures and technologies involved, the large volumes of radioactive wastes, and chronically understated nuclear construction costs.

Research done at the Rand Corporation introduces another element of uncertainty. Analysts there have concluded that large-scale engineering projects dependent on newly developed technologies cost on average four times more than predicted at the outset. Recent U.S. nuclear power plant construction costs have amounted to five to ten times the original estimate, even after accounting for inflation. Cost overruns of several hundred percent have become the norm rather than the exception.

Because of gross cost overruns, partially constructed nuclear plants worth approximately $20 billion have been abandoned in the United States. In the early seventies, reactors were projected to come into service at less than $1 million per megawatt, but none has done so in recent years. Costs for just-completed units have averaged closer to $3 million per megawatt and at the not-yet-operating Shoreham plant on Long Island, the figure has already surpassed $5 million per megawatt.

Decommissioning costs for two of the first generation of retired reactors are estimated to be more than $1 million per megawatt. The Shippingport facility, despite a unique transportation arrangement and federally subsidized waste disposal, is expected to cost $98 million to decommission—$1.36 million per megawatt. This figure excludes the cost of dismantling noncontaminated buildings.

It is debatable whether these estimates can be directly scaled on a per megawatt basis. Yet the amount of radioactivity that builds up in a plant is proportional to the plant's capacity multiplied by the number of years it operates, although economies of scale will be involved. If costs are scaled according to reactor size, then the Shippingport experience—despite the shortsighted savings being realized by keeping the reactor pressure vessel intact during transport—indicates that decommissioning one of today's large reactors could cost over $1 billion.

André Cregut of the French atomic energy commission estimates that utilities will not decommission plants until the cost can be brought down to about 15 percent of initial investment, compared with the 40 percent it is likely to cost using currently available techniques. At 40 percent of investment, decommissioning costs would approach $1 billion for the plants that have recently gone into operation. Whether the expense can be lowered is the crucial, but unanswered, question.

Because nuclear reactor designs have changed so much over the years, very few countries—perhaps only France and Canada with their uniform construction programs—will be able to design a uniform decommissioning system. Nine different designs are represented among the 20 power reactors that have shut down. Decommissioning cost estimates and experience gained may not be transferable among utilities.

One of the largest components of decommissioning costs will be waste disposal—up to 40 percent of the total, according to some industry analysts. In the last decade, the cost of shallow land burial of a 55-gallon drum of low-level wastes has increased more than tenfold in the United States. Waste disposal costs have tripled in the last five years and are likely to increase steadily. Current disposal facilities will continue to raise their rates, and the start-up costs of new sites for both low- and high-level wastes are likely to be high.

Labor expenses may also rise as efforts are made to minimize the radiation doses that personnel receive. Replacement of steam generators at the Surry reactor in Virginia, for example, required more than three times the expected worker hours. Current regulations limit exposure of nuclear industry workers to 5 rems per year, 10 times as much as permitted for an average person. Many health experts would like the ceiling to be lowered, a move that would significantly increase labor costs during decommissioning.

The number of years a reactor remains in storage before dismantlement is a variable frequently overlooked when formulating decommissioning cost estimates. Providing on-site surveillance for several decades is expensive. The bill will include salaries for a skeleton staff, radiation monitoring equipment, and, if a high-level waste repository is not available, maintenance of fuel-storage facilities. Postponement of decommissioning will also result in the loss of staff most familiar with the plant and require excellent record keeping to inform the future crew of the reactor's intricacies and its operating history.

As experience is gained in decommissioning and waste handling, regulations are likely to become more strict. If worker radiation exposure limits are lowered, if residual radioactivity standards are set at levels more stringent than predicted, or if transportation and disposal rules are tightened, costs could rise substantially.

Paying for Decommissioning

Ensuring that adequate funds will be available when needed for decommissioning is a task that utility executives, regulatory commissioners, and politicians have for the most part chosen to ignore. But unless money is set aside during the years the plant is operating, the bill would be charged to those who did not use the power, leading to solvency problems for utilities, or resulting in short cuts in the safety measures for decommissioning.

Electric utilities often argue that money collected for decommissioning should be placed in the companies' general revenues and used for all types of expenses, including the construction of more nuclear plants. This strategy allows utilities to bypass capital markets and keep debt payments low, but the method has been termed "phantom funding" by some observers because the money is invested in assets and is not readily available to the utility. In a study conducted for NRC, University of Pennsylvania economist J. J. Siegel concluded that if an outside decommissioning fund were established, "it would be virtually impossible for the utility to divert these assets for other uses, and funds would be assured no matter what events, legal or financial, occur."

Of the four retired commercial reactors in the United States, only one unit began collecting decommissioning funds before it shut down. The Pacific Gas & Electric Company, owner of the Humboldt Bay reactor on the northern California coast, collected $500,000 during the plant's last four years of operation. The lack of financial planning in three instances, and the late and inadequate implementation in the fourth, have led eight states (California, Colorado, Maine, Massachusetts, Mississippi, New Hampshire, Pennsylvania, and Vermont) to require mandatory periodic deposits into external savings accounts.

In Sweden, each reactor operator pays an annual fee to the government. The funds are invested in separate accounts from which the utilities can borrow money to pay for decommissioning. Less formal arrangements exist in West Germany and Switzerland. In France, the "when-needed" principle of funding has been adopted. Utility managers assume they will be able to request funds from the national treasury when the time comes to decommission their reactors. The French utility debt is already $30 billion because of the accelerated nuclear construction program, and decommissioning costs may result in large government budget deficits.

Regardless of the role of nuclear power in a nation's energy plans, existing plans must eventually be scrapped. Decommissioning bills will first fall due in those countries that pioneered the development of this energy source. And their skill at managing the expense will be closely watched. Nations with newer reactors will learn valuable lessons, and countries that have not yet built such plants will be better able to assess the true lifetime costs of nuclear energy.

Paid over a period of thirty years by all the customers of a utility, decommissioning bills are affordable. The average residential electricity consumer in the United States, if served by a completely nuclear electric utility, would have been charged an annual fee of about $55 in 1984 for decommissioning. (This assumes a total decommissioning bill of $1 million per megawatt of plant capacity.) Customers with inefficient electric heating and air conditioning systems might have to pay well over $100 a year, but those using electricity efficiently would pay far less than the average. And the actual charges would be even lower, because no U.S. electric utility relies exclusively on nuclear power.

A Long-Term Strategy

During the next three decades, over three hundred nuclear power plants will be shut down. Eventually, the international nuclear community will have to dispose of more than five hundred plants, including those currently under construction. Some of the reactors that have already been retired are large enough and have operated long enough to yield valuable lessons for future decommissioning projects.

Decommissioning planning has lagged far behind reactor development. The International Atomic Energy Agency (the United Nations research and watchdog arm of the nuclear industry) did not hold its first meeting on decommissioning until 1973, some nineteen years after the first power reactor was built. The initial technical meeting sponsored by the agency was not convened until two years later. This neglect of the back end of the nuclear fuel cycle has been replicated by national atomic energy authorities everywhere.

The biggest stumbling block for all nations with nuclear plants is the lack of permanent disposal facilities for radioactive wastes. Although many reactor operators around the world agree that plants should be dismantled as quickly as practicable after shutdown, that option has been foreclosed until at least the turn of the century for lack of disposal sites. No country currently has the capability to permanently dispose of the high-level wastes now stored at a single reactor.

Detailed technical guidelines for decommissioning should be expedited so that utilities can plan for the future. Early knowledge of decommissioning requirements would also allow engineers to incorporate design changes that would facilitate later decontamination and dismantlement. A simple concept that eluded manufacturers of the first nuclear plants was the value of putting a protective coating on all surfaces that would be exposed to radiation. Even a thick layer of removable paint reduces the surface contamination of structural components. Limited experience with neutron-activated wastes has also demonstrated the need to minimize the amount of neutron-absorbing impurities used in reactor steel and concrete.

Knowledge of what can be done to make the decommissioning process safer and less costly is slowly accumulating, but efforts to expand that information base need to be strengthened. The temptation to use cost-cutting measures in the first projects should be resisted. Saving millions of dollars now could mean spending billions of extra dollars later.

BIBLIOGRAPHY

An asterisk (*) preceding a reference indicates that the article or part of it has been reprinted in this book.

BOOKS AND PAMPHLETS

*American Nuclear Society. Nuclear energy facts: questions and answers. American Nuclear Society. '85.

Bartlett, Donald and Steele, James. Forevermore, nuclear waste in America. W. W. Norton. '85.

Brown, Kirk W. Hazardous waste land treatment. Butterworth. '83.

Burns, Michael E. Low-level radioactive waste regulation. Lewis. '88.

Colglazier, E. W. The politics of nuclear waste. Pergamon Press. '82.

Dlouhy, Zdenek. Disposal of radioactive waste. Elsevier. '82.

Duedall, Iver W. Energy wastes in the ocean. Wiley. '85.

Epstein, Samuel S. and Brown, Lester O. Hazardous waste in America. Sierra Club Books. '82.

Goldman, Benjamin A. Hazardous waste management: reducing the risk. Island Press. '86.

Grisham, Joe W. Health aspects of the disposal of waste chemicals. Pergamon. '86.

Grossman, Karl. The poison conspiracy. Permanent Press. '83.

Johansson, Thomas B. and Steen, Peter. Radioactive waste from nuclear power plants. University of California Press. '81.

Kirschman, Don. To burn or not to burn: the economic advantages of recycling over garbage incineration for New York City. Environmental Defense Fund. '85.

Krauskopf, Konrad. Radioactive waste disposal and geology. Chapman and Hall. '88.

Lau, Foo-Sun. Radioactivity and nuclear waste disposal. Wiley. '87.

League of Women Voters. The nuclear waste primer. N. Lyons. '85.

Lester, James P. and McBowman, Ann O., eds. The politics of hazardous waste management. Duke University Press. '83.

Lowrance, W. W. Public health risk of the dioxins. Rockefeller University. '84.

Murray, Raymond. Understanding radioactive waste. Battelle Press. '82.

National Research Council. Social and economic aspects of radioactive waste disposal. National Academy Press. '84.

Pishdadazar, H. and Mighissi, A. Alan. Hazardous waste sites in the United States. Pergamon Press. '81.

Pollock, Cynthia. Mining urban wastes: the potential for recycling. Worldwatch Institute. '87.

Shapiro, Fred C. Radwaste. Random House. '81.

Squires, Donald F. The ocean dumping quandary. State University of New York. '83.

U.S. Government. Ocean incineration: its role in managing hazardous waste. Congress of the U.S., Office of Technology Assessment. '86.

Vogler, John. Work from waste: recycling wastes to create employment. Oxfam. '81.

Walker, Charles et al. Too hot to handle? Yale University Press. '83.

PERIODICALS

*Where will all the garbage go? MacFadyen, J. Tevere. Atlantic. 255:29–33. Mr. '85.

The toxic tornado. Berle, P. A. A. Audubon. 87:4. N. '85.

*Fecal follies. Wolff, A. Audubon. 88:32+. Ja. '86.

Storing of the world's spent nuclear fuel. Barkenhus, J. N. Bulletin of the Atomic Scientists. 41:34–37. N. '85.

Problems with military nuclear waste. Lawless, W. F. Bulletin of the Atomic Scientists. 41:38–42. N. '85.

A drug giant plagued by dioxin's poison. Business Week. 42–43. My. 2, '83.

The tug of war that's tying up Superfund. Recio, M. E. Business Week. p. 45. O. 28, '85.

Everybody will probably pay to clean up toxic waste. Recio, M. E. Business Week. p. 30. N. 4, '85.

The Superfund bill puts Reagan in a bind. Hoppe, R. M. Business Week. p. 30. O. 20, '86.

Garbage: it isn't the other guy's problem anymore. Thompson, T. and Bluestone, M. Business Week. pp. 50–51+. My. 25, '87.

*Troubled waters. Smart, T. and Smith, E. T. Business Week. pp. 88–91, 94, 98, 102, 104. O. 12, '87.

Superfund controversy. Congressional Digest. 65:163–92. Je./Jl. '86.

Hazardous waste: a health hazard for our wild life. North, J. Conservationist. 89:42–47. Ja./F. '85.

Cleanup offers a second chance. Kadluck, M. Conservationist. 39:32–35. Mr./Ap. '85.

Recycling—not a lot of garbage. Kaplan, E. Current Health. 12:28–29. D. '85.

Ocean dumping in the New York bight. Payton, Beverly M. Environment. 27:26–32. N. '85.

*The year the bight died. Payton, Beverly M. Environment. 27:29. N. '85.

Low-level radioactive waste policy. Dineen, J. Environment. 27:35–36. D. '85.

*The closing act: decommissioning nuclear power plants. Pollock, Cynthia. Environment. 28:10–15+. Mr. '86.

Hazardous wastes. Environment. 28:2–20+. Ap. '86.

Industrial waste reduction: the process problem. Valentino, F. W. and Walmet, G. E. Environment. 28:16–20+. S. '86.

Waste reduction: a new strategy to avoid pollution. Oldenburg, K. U. and Hirschhorn, J. S. Environment. 29:16–20+. Mr. '87.

Laying nuclear waste to rest. Environment. 6–20+. O. '87.

*Risky business. Cook, J. Forbes. 136:106–7+. D. 2, '85.

Cleaning up after industry's slobs. Pomice, Eva. Forbes. 139:90. Ap. 20, '87.

Dow vs. the dioxin monster. Main, Jeremy. Fortune. 107:82–86+. My. 30, '83.

No more wasteland. Fortune. 111:9–10. Ja. 21, '85.

The sweet smell of profits from trash. Leinster, C. Fortune. 111:150–54. Ap. 1, '85.

Contaminated water. Foegen, J. H. Futurist. 20:22–24. Mr./Ap. '86.

Radioactive headache. Futurist. 20:47–48. Mr./Ap. '86.

Hazardous waste: prevention and cure. Futurist. 21:47. Mr./Ap. '87.

The economics of garbage. Futurist. 21:42–43. N./D. '87.

*Bad news from Britain. Robinson, M. Harper's. 270:65–72. F. '85.

The United States toxic deadlock. Austen, I. Macleans. 98:21. Ap. 29, '85.

Cleaning up a poisonous spill. Rose, M. Macleans. 98:52. My. 6, '85.

A town that demands answers. McAndrews, B. Macleans. 98:16. Ag. 5, '85.

Covering up deadly waste. Finlayson, A. Macleans. 99:35. My. 12, '86.

Is your town a Love Canal? Smith, E. Mother Earth News. 96:110–12. N./D. '85.

New ways to tackle toxic wastes. Garland, A. W. and Sinclair, M. Ms. 13:135–38. O. 3, '87.

*The export of U.S. toxic wastes. Porterfield, A. and Weir, D. Nation. 245:325+. O. 3, '87.

Fighting the nuke-waste shell game. Russell, D. Nation. 245:577+. N. 21, '87.

Storing up trouble . . . hazardous waste. Boraiko, A. A. National Geographic. 167:318-51. Mr. '85.

Putting the heat on polluters. Stranahan, S. O. National Wildlife. 23:30-33. Ag./S. '85.

Coming soon: a move to tackle hazardous waste. Hair, J. D. National Wildlife. 23:30. O./N. '85.

New teeth in waste law. Nation's Business. 74:16. N. '86.

Six ounces of gasoline in a soda can. Chandler, W. U. Natural History. 94:79. Ap. '85.

What do you do with a worn-out nuclear reactor? Pollock, Cynthia. Natural History. 95:61-64. Ap. '86.

How toxic is dioxin? the mystery deepens. Joyce, Christopher. New Scientist. 110:24. My. 1, '86.

The disposal alternative. Browne, M. W. New York Times Magazine. p. 24. My. 11, '86.

The trouble at Times Beach. Lerner, Michael A. Newsweek. 101:24. Ja. 10, '83.

Toxic wastes: the dioxin scare spreads. Newsweek. 101:28. My. 23, '83.

Dioxin: how great a threat? Begley, Sharon. Newsweek. 102:65-66. Jl. 11, '83.

America's toxic tremors. Starr, M. Newsweek. 106:18-19. Ag. 26, '85.

Rubbish on the high seas. Osterberg, C. Newsweek. 106:18. O. 7, '85.

A nuclear burial ground. Beck, M. Newsweek. 107:30-31. Je. 16, '86.

Ohio's toxic nightmare. Newsweek. pp. 108-19. Jl. 21, '86.

Our befouled beaches. Adler, Jerry. Newsweek. 110:50-51. Jl. 27, '87.

*Blight in the bight: sewage and water don't mix. Payton, B. M. Oceans. 18:63-67. My./Je. '85.

Society challenges incineration. Oceans. 18:67. S./O. '85.

Radwaste dumping. Oceans. 18:68. N./D. '85.

Worrisome ocean wastes. Oceans. 19:5. S./O. '86.

OTA issues warning. Oceans. 20:5. Jl./Ag. '87.

*Ocean dumping. Farrington, John W. et al. Oceanus. 25:39-50. Winter '82/'83.

A manager who blew the whistle. Snel, A. Progressive. 49:17. Ap. '85.

The Superfund stench. Nanetick, M. Progressive. 49:18-19. N. '85.

Is Maine Yankee headed for the sump? Kriesberg, J. Progressive. 51:12-13. N. '87.

Chemicals from waste dumps. Abelson, Philip H. Science. 229:335. Jl. 26, '85.

DOE, states reheat nuclear waste debate. Crawford, Mark. Science. 230:150-51. O. 11, '85.

Treatment of hazardous wastes. Abelson, Philip H. Science. 233:509. Ag. 1, '86.

Nuclear waste program faces political burial. Marshall, Eliot. Science. 233:835-36. Ag. 22, '86.

OTA enters inflamed debate on ocean incineration. Sun, Marjorie. Science. 233:934. Ag. 29, '86.

OTA urges waste reduction as dump sites close. Crawford, Mark. Science. 233:1381. S. 26, '86.

Transportation of hazardous materials. Abelson, Philip H. Science. 234:125. O. 10, '86.

Hazardous waste: where to put it? Crawford, Mark. Science. 235:156-57. Ja. 9, '87.

Discovering microbes with a taste for PCBs. Roberts, L. Science. 237:975-77. Ag. 28, '87.

Nevada wins the nuclear waste lottery. Marshall, E. Science. 239:15. Ja. 1, '88.

*Burying nuclear waste. Yulsman, Tom. Science Digest. 93:16. Jl. '85.

*The deadliest garbage of all. Stranahan, Susan Q. Science Digest. 94:64+. Ap. '86.

Finding a resting place for radwaste. Peterson, I. Science News. 127:6. Ja. 5, '85.

Incineration on the high seas. Matthewson, J. Science News. 127:406. Je. 29, '85.

Dioxin: is everyone contaminated? Raloff, Janet. Science News. 128:26-29. Jl. 13, '85.

House passes tough Superfund bill. Peterson, I. Science News. 128:390. D. 21-28, '85.

Radwaste crisis narrowly averted. Raloff, Janet. Science News. 129:22-23. Ja. 11, '86.

Retiring reactors: what's the cost? Raloff, Janet. Science News. 129:230. Ap. 12, '86.

EPA cancels ocean incineration. Science News. 129:395. Je. 21, '86.

Absolutely swimming in bacteria. Science News. 130:26. Jl. 12, '86.

Five-fold increase in Superfund money. Murray, M. Science News. 130:86-87. Ag. 9, '86.

OTA warning on wastes in coastal waters. Weisburd, S. Science News. 131:309. My. 16, '87.

*Dioxin. Tschirley, Fred H. Scientific American. 254:29-35. F. '86.

Year of the Superfund. Green, R. Sierra. 70:36+. My./Je. '85.

*Low-level lowdown. Warner, G. Sierra. 70:19-20+. Jl./Ag. '85.

Cities rush to recycle. Barry, P. Sierra. 70:32-35. N./D. '85.

*Biology's answer to toxic dumps. Holbrook, J. J. Sierra. 72:24–25+. Ja./F. '87.

Low-level nuclear waste: who makes it, who will take it. Stine, A. Sierra. 72:16–17. S./O. '87.

*Garbage in, garbage out. Mann, Carolyn. Sierra. 72:20+. S./O. '87.

*The case for ocean waste disposal. Lahey, William and Connor, Michael. Technology Review. 86:60–68. Ag./S. '83.

Burning wastes at sea. Barnett, Robert. Technology Review. 87:85+. F./Mr. '84.

A common-sense approach to nuclear waste. Gilinsky, Victor. Technology Review. 88:14+. Ja. '85.

The missing links: restructuring hazardous-waste controls in America. Piasecki, B. and Gravander, J. Technology Review. 88:42–49+. O. '85.

Radioactive brinkmanship. Mattill, J. I. Technology Review. 88:78+. N./D. '85.

*Garbage: to burn or not to burn? Davis, T. Technology Review. 90:19. F./Mr. '87.

*Burning trash: how it could work. Hershkowitz, Allen. Technology Review. 90:26–34. Jl. '87.

*Recycling: coming of age. Goldoftas, Barbara. Technology Review. 90:28–35+. N./D. '87.

*A problem that cannot be buried. Magnuson, Ed. Time. 126:76–78+. O. 14, '85.

Waste dump wanted. Time. 130:70. Jl. 20, '87.

Give me your wretched refuse. Trippett, F. Time. 130:95. N. 23, '87.

Cleanup of toxic-waste dumps. U.S. News & World Report. 98:48. F. 25, '85.

$1 billion later, toxic cleanup barely begun. Taylor, R. A. U.S. News & World Report. 98:57–58. Ap. 22, '85.

Public is willing to pay price for pure air and water. U.S. News & World Report. 99:41–42. Jl. 8, '85.

Toxic cleanup: a prisoner of politics on Capitol Hill. Walsh, K. T. and Taylor, R. A. U.S. News & World Report. 99:32. N. 25, '85.

Why pollution watchdogs can't bite. Taylor, R. A. U.S. News & World Report. 100:76. F. 10, '86.

West digs in to fight its own nuclear war. Duth, B. U.S. News & World Report. 101:18–19. Ag. 4, '86.

Tons and tons of trash and no place to put it. Budiansky, Stephen. U.S. News & World Report. 103:58–62. D. 14, '87.

*The midnight dumpers. Miller, Judith and Miller, Mark. USA Today. 113:60–64. Mr. '85.

Coastal ocean toxic waste pollution: where are we and where do we go?
Kao, T. W. and Bishop, J. M. USA Today. 114:20–23. Jl. '85.